London Mathematical Society Lecture Note Series. 2

F. F. BONSALL and J. DUNCAN

Numerical Ranges
of Operators on Normed Spaces
and of Elements of Normed Algebras

CAMBRIDGE AT THE UNIVERSITY PRESS 1971

Published by The Syndics of the Cambridge University Press

Bentley House, 200 Euston Road, London N. W. 1

American Branch: 32 East 57th Street, New York, N. Y. 10022

Library of Congress Catalogue Card No. : 71-128498

ISBN: 0 521 07988 8

Printed offset in Great Britain by
Alden & Mowbray Ltd at the Alden Press, Oxford

Contents

Page

Introduction   1

CHAPTER 1

Numerical range in unital normed algebras   13

2.   Elementary theorems   14

3.   The exponential function   26

4.   Numerical radius theorems   33

CHAPTER 2

Hermitian elements of a complex unital Banach

algebra   45

5.   Vidav's lemma and applications   46

6.   The Vidav-Palmer theorem   56

7.   Applications of the Vidav-Palmer theorem to

B*-algebras   67

8.   Other applications of the Vidav-Palmer

theorem   73

CHAPTER 3

Operators   80

9.   The spatial numerical range   81

10.   Spectral properties   88

11.   Geometrical and topological properties

of V(T)   98

CHAPTER 4

Some recent developments   105

12.   The second dual of a Banach algebra   106

13.   Spectral states   111

14.   Remarks and problems   123

BIBLIOGRAPHY   131

INDEX   138

Introduction

The numerical range of a linear operator on a normed
linear space is a subset of the scalar field constructed in such a
way that it is related both to the algebraic and the norm structures
of the operator. In this it differs from the spectrum, which is
related to the algebraic structure but independent of the norm (up
to equivalence). For an operator on a Hilbert (or pre-Hilbert)
space the numerical range has a very natural definition which was
introduced, for finite dimensional spaces, by Toeplitz in 1918
[70], as follows. Let $\mathcal{H}$ denote a pre-Hilbert space with scalar
product $<,>$ and norm $\|\cdot\|$, and let $S(\mathcal{H})$ denote the unit
sphere in $\mathcal{H}$, $S(\mathcal{H}) = \{x \in \mathcal{H} : \|x\| = 1\}$. Then the numerical
range of a linear operator $T: \mathcal{H} \to \mathcal{H}$ is the set $W(T)$ of scalars
defined by

$$W(T) = \{<Tx, x> : x \in S(\mathcal{H})\} . \qquad (1)$$

The numerical range $W(T)$ owes part of its motivation to the
classical theory of quadratic forms, but modern developments are
naturally given in terms of the theory of bounded linear operators.

We have not included an account of the numerical range of
an operator on a Hilbert space in these notes. There are two
reasons for this. First and most important is the ready availability
of an excellent account of the subject in Halmos's recent book [30].

The second reason is that in the context of Hilbert spaces, with the varied and powerful methods available there, the numerical range has remained a relatively unimportant tool. By comparison, the theory of operators on Banach spaces and the general theory of Banach algebras are lacking in effective methods, and therefore such methods as we have are to be correspondingly treasured. Moreover the numerical range can be applied to problems, such as the metric characterization of B*-algebras, which become meaningful only in a wider context, even though, as in this instance, they may concern a Hilbert structure.

It is, however, appropriate to give a brief review of the main results for Hilbert spaces, since these have motivated much of the general theory. It is obvious that $W(T)$ contains all eigenvalues of T; for if λ is an eigenvalue, there exists $u \in S(\mathcal{H})$ with $Tu = \lambda u$, and then

$$\lambda = \lambda <u, u> \; = \; < Tu, u> \; \in W(T) \; .$$

Thus when $\mathcal{H}$ has finite dimension, $W(T)$ contains $Sp(T)$, the spectrum of T, and Toeplitz proved that the unbounded component of the complement of $W(T)$ has a convex curve for its boundary. The truth of this latter result became obvious when it was proved by Hausdorff [31] , still for finite dimensional $\mathcal{H}$, that $W(T)$ is convex. This last result of Hausdorff was extended by Stone [68] to operators on pre-Hilbert spaces of arbitrary dimension. Let $\mathcal{H}$ denote a complex Hilbert space, let T be a bounded linear operator on $\mathcal{H}$ with operator norm $|T|$, and let

$$w(T) = \sup \{ \, |\zeta| : \zeta \in W(T) \, \} \; .$$

2

Further developments in the theory included the results that $Sp(T) \subset W(T)^-$ (the closure of $W(T)$), that, for a normal operator T, $W(T)^-$ is the convex hull of $Sp(T)$, and that for all T

$$|T| \leq 2 w(T).$$
(2)

These elementary results already give examples of the relationship of the numerical range to both the algebraic and the norm properties of the operator.

The study of the numerical range of an operator on a Hilbert space continued in an unspectacular fashion until quite recently. One very interesting recent result is the theorem of Berger [4] that

$$w(T^n) \leq (w(T))^n \quad (n = 1, 2, \dots),$$
(3)

which is remarkable because the inequality $w(ST) \leq w(S)w(T)$ is false, as is also the special case $w(T^{n+m}) \leq w(T^n)w(T^m)$. An elegant and simple proof of the inequality (3) is given by Pearcy [54]. For other references see Halmos [30] and Putnam [56].

By contrast to the long history of the Hilbert space numerical range, the birth of the general theory was long delayed and its growth has been spectacular. No concept of numerical range appropriate to general normed linear spaces appeared until 1961 and 1962, when distinct, though related, concepts were introduced independently by Lumer [40] and Bauer [3]. Lumer defined the concept of a semi-inner-product on a linear space, and showed that every normed linear space $(X, \|.\|)$ has at least one semi-inner-product $[,]$ such that

$$[x, x] = \|x\|^2 \quad (x \in X).$$
(4)

3

In terms of a semi-inner-product satisfying (4), the definition (1) used for Hilbert spaces at once generalizes to give the definition of the numerical range W(T) for a linear operator on **X**,

$$W(T) = \{[Tx, x] : x \in S(X)\},$$

where S(X) denotes the unit sphere of **X**. On the face of it this definition has the serious defect that it is not an invariant of the normed space (**X**, $\|\cdot\|$), since, except when the unit ball of **X** is smooth, there are infinitely many semi-inner-products on **X** satisfying (4). However, this defect is more apparent than real, for Lumer proved that

$$\sup\{\operatorname{Re}\lambda : \lambda \in W(T)\} = \lim_{\alpha \to 0+} \frac{1}{\alpha}\{|I + \alpha T| - 1\}, \qquad (5)$$

from which it follows that $\overline{co}\ W(T)$, the closed convex hull of W(T), is independent of the choice of semi-inner-product satisfying (4). In fact, (5) shows that $\overline{co}\ W(T)$ depends only on the norms of the operators in the two dimensional linear subspace spanned by I (the identity operator) and T.

Lumer's paper [40] was undoubtedly the most important in the development of the subject. Besides introducing a suitable general concept and obtaining its fundamental properties, including a generalization of the crucial inequality (2), this paper showed the power of the concept in applications both to linear operators and to Banach algebras. Given an element a of a normed algebra **A**, let T_a denote the left regular representation operator:

$$T_a x = ax \qquad (x \in A). \qquad (6)$$

Then the numerical range $W(a)$ may be defined by $W(a) = W(T_a)$ in terms of some semi-inner-product on A related to the norm of A. Lumer showed that the numerical range is an effective tool for relating algebraic and geometric properties of a Banach algebra, giving, in particular, a simple proof of the theorem of Bohnenblust and Karlin [10] that the unit element is a vertex of a complex unital Banach algebra, and throwing fresh light on Vidav's metric characterization of B*-algebras [72].

Bauer's paper [3] was concerned only with finite dimensional normed linear spaces, but the concept of numerical range that he introduced is available without restriction of the dimension. Let $(X, \|.\|)$ be a normed linear space, $S(X)$ its unit sphere, X' its dual space, and let

$$\Pi = \{(x, f) \in S(X) \times S(X'): f(x) = 1\} .$$

Then, for an operator T on X, the numerical range $V(T)$ is defined by

$$V(T) = \{f(Tx): (x, f) \in \Pi\} . \tag{7}$$

When X is a Hilbert space, $(x, f) \in \Pi$ if and only if $x \in S(X)$ and f is the functional given by $f(y) = \langle y, x \rangle$ $(y \in X)$. Thus $V(T)$ in this case coincides with the classical $W(T)$ given by (1) above. If X is a normed linear space with a smooth unit ball, then $V(T)$ coincides with the numerical range $W(T)$ corresponding to the unique semi-inner-product satisfying (4). For a general normed linear space X, $V(T)$ is the union of all the numerical ranges $W(T)$ corresponding to all choices of semi-inner-product satisfying (4), and for each choice of such a semi-inner-product

5

$$\overline{co} \ V(T) = \overline{co} \ W(T) \ .$$

Thus Lumer's results are immediately transferable to $V(T)$.

Since $V(T)$ is intrinsically defined, without intervention of a choice of semi-inner-product, it is to be expected that it will exhibit greater regularity than $W(T)$; and two recent theorems exemplify this. Williams [74] has proved that for any bounded linear operator T on a complex Banach space

$$Sp(T) \subset V(T)^- \ ,$$

whereas we only know that $W(T)^-$ contains the approximate point spectrum of T; and Bonsall, Cain, and Schneider [12] have proved that for every bounded linear operator T on a normed linear space $V(T)$ is connected, whereas $W(T)$ may be disconnected. With one trivial exception, the connectedness of $V(T)$ holds for any continuous mapping T of $S(X)$ into X, and this suggests that $V(T)$ may be useful for the study of non-linear mappings. It should be remarked that $V(T)$ is not in general convex, and perhaps the failure of the best known property of the Hilbert space numerical range contributed to the long delay in the introduction of $V(T)$, which after all is a very simple generalization of the Hilbert space numerical range.

That the numerical range can give interesting information even about an $n \times n$ matrix is exemplified by the following theorem of Nirschl and Schneider [47]. Let λ be an eigenvalue of an operator T which belongs to the frontier of co $V(T)$; then λ has index (ascent) one, i. e.

$$(\lambda I - T)^2 x = 0 \implies (\lambda I - T)x = 0 \ .$$

6

This result is immediately applicable to the eigenvalues of modulus one of a stochastic matrix.

Let $v(T) = \sup\{|\lambda| : \lambda \in V(T)\}$. Since $\overline{co}\,V(T) = \overline{co}\,W(T)$, we have $v(T) = \sup\{|\lambda| : \lambda \in W(T)\}$ for any choice of semi-inner-product satisfying (4). For complex normed linear spaces, we have

$$|T| \le e\,v(T),$$

an inequality essentially due to Bohnenblust and Karlin [10], in which the constant $e\ (= \exp 1)$ has been shown by Glickfeld [28] to be best possible for the class of all complex normed linear spaces. For special classes of spaces better constants have been established. The reciprocal of the least constant valid for a given normed linear space is called the numerical index of the space. It has recently been proved by Duncan, McGregor, Pryce and White [21] that for every real number $\nu \in [e^{-1}, 1]$ there exists a complex normed linear space with numerical index ν.

The inequality (3) has not been established[†] for operators on a general Banach space (perhaps for the natural reason), but M. J. Crabb [17] has recently proved that if T is a bounded linear operator on a complex Banach space normalized so that $v(T) \le 1$, then

$$\|T^n\| \le e\,n^{\frac{1}{2}} \quad (n = 1, 2, \dots).$$

He has further proved, using the theorem of Nirschl and Schneider mentioned earlier, that if X is finite dimensional (or more

---

† See Remark (10) of §14.

7

generally if T is a meromorphic operator), then with the same normalization the sequence $\{ \| T^n \| \}$ is bounded.

Suppose now that A is a normed algebra, and is unital, i.e. has a unit element 1 with $\| 1 \| = 1$. Given $a \in A$, the numerical range $V(a)$ is defined by $V(a) = V(T_a)$, where T_a is the left regular representation operator on A, as in (6) above. In this case, however, a remarkably simple expression is available for $V(a)$. Let $D(1)$ denote the set of all normalized states on A, i.e. continuous linear functionals f on A such that

$$f(1) = \| f \| = 1 .$$

Then

$$V(a) = \{ f(a) : f \in D(1) \} .$$

It follows at once that $V(a)$ is a compact convex set; and a number of other fundamental properties of $V(a)$ are very easily established.

Given a bounded linear operator T on a normed linear space X, we can regard T as an element of the unital normed algebra $B(X)$ of all bounded linear operators on X, and we then have two intrinsic numerical ranges available for T. It turns out, as would be expected, that the numerical range of T as an element of the normed algebra $B(X)$ is the closed convex hull of the spatial numerical range $V(T)$ given by (7) above.

Of special interest are the Hermitian elements of a complex unital Banach algebra A, i.e. elements h of A such that $V(h) \subset \underset{\sim}{R}$. By the lemma of Lumer mentioned above (5), h is Hermitian if and only if

8

$$\lim_{\alpha \to 0} \alpha^{-1} \{ \| 1 + i\alpha h \| - 1 \} = 0 ,$$

i. e. if and only if h is Hermitian in the sense of Vidav [72]. Thus a fundamental lemma of Vidav shows that, for all Hermitian elements h of A, we have

$$\text{co Sp(h)} = V(h) . \qquad (8)$$

This is a crucial lemma in the Vidav characterization of B*-algebras, already mentioned above, which, as recently improved by Palmer [49], states that A is a B*-algebra if and only if every element of A is of the form h + ik with h and k Hermitian. As is well known, B*-algebras are isometrically star isomorphic to C*-algebras, i. e. uniformly closed self-adjoint algebras of operators on Hilbert spaces. Thus an important theorem about algebras of operators on Hilbert spaces involves the concept of numerical range for general normed linear spaces and not just the classical Hilbert space concept. To show the effectiveness of the Vidav-Palmer theorem, we give a number of applications of the theorem. In particular we use it to give new and transparent proofs of the theorem of Glimm and Kadison on a Banach star algebra satisfying the condition

$$\| a^*a \| = \| a^* \| \, \| a \| ,$$

and of the well known theorems of Kaplansky and of Sherman that the quotient of a B*-algebra by a closed two-sided ideal and the second dual of a B*-algebra with the Arens multiplication are both B*-algebras.

It was proved by Lumer [42] that there exist Hermitian elements h such that h^2 is not Hermitian, and we give a very simple example of such an element h due to M. J. Crabb (unpublished). It follows that a Hermitian element h may fail to generate a B*-subalgebra of A. It is therefore very remarkable that Sinclair [62] has been able to strengthen the equality $v(h) = \rho(h)$, which is equivalent to Vidav's lemma (8), to

$$\|h\| = \rho(h) ,$$

where $\rho(h)$ denotes the spectral radius of h.

The theory of the numerical range may be regarded as part of the study of the geometry of Banach algebras A, and specifically of their dual spaces A'. Certain elements of A', the multiplicative linear functionals on commutative algebras A and the positive linear functionals on star algebras A have been central to the theory of Banach algebras from the beginning. More recently Fell [22] and Bonsall and Duncan [13] have studied the dual representations of A arising from general elements of A'. The general study of A' is beyond the limits of the present notes, but we have regarded a brief study of the spectral states as coming within them because of their special connection with the numerical range.

We are indebted to R. T. Moore for drawing our attention to the interest of those elements of A' that are normalized states of A for every choice of algebra-norm on A equivalent to the given norm. Such f are characterized by the property

$$f(a) \in co\ Sp(a) \qquad (a \in A) ,$$

10

and we therefore call them spectral states. They have trace-like
properties, in particular

$$f(ab) = f(ba) \qquad (a, b \in A) \ ,$$

from which it follows that the algebra $B(\mathcal{H})$ for an infinite
dimensional Hilbert space $\mathcal{H}$ does not have any spectral states.
The algebra of all $n \times n$ matrices has exactly one spectral state,
namely the normalized trace; and, at the other extreme, every
probability measure on the carrier space of a commutative unital
Banach algebra gives rise to a spectral state on A.

In organizing these notes, we have chosen to reverse the
historical order and treat first the numerical range of a unital
normed algebra, since this gives the smoothest and most satisfactory theory and appears to cover the bulk of applications. We then
more briefly develop those properties of the numerical range of an
operator in the senses of Lumer and Bauer that are not covered
by the unital normed algebra theory. We end with a few recent
developments and loose ends.

These notes are intended to be expository in nature, and
full proofs are given, based only on standard and more or less
elementary material on normed linear spaces and normed algebras,
except that a few side remarks are made without proofs and at one
point in §6 we make use of a theorem on C*-algebras.

There is little in these notes other than applications and
new proofs for which we personally can claim originality, but,
since the subject is developing very rapidly at the present time,
we have judged that the usefulness of the notes would be increased
by including remarks on so far unpublished results of which we
have knowledge through preprints, letters or conversation. We

are particularly grateful to M. J. Crabb, G. Lumer, R. T. Moore, T. W. Palmer, A. L. T. Paterson, A. M. Sinclair and J. P. Williams for keeping us informed and for allowing us to mention their ideas and results.

The second author gave a lecture course at Aberdeen based on the material in chapters 1-3 and is grateful to C. M. McGregor and S. McKilligan for making a careful record of the lectures. We are indebted to Miss Marjory McKinnon for careful typing of the entire manuscript. Finally we are grateful to the London Mathematical Society for this opportunity to present the theory of numerical range.

1. Numerical Range in Unital Normed Algebras

In a Banach algebra the spectrum of an element depends only on the algebra structure and not at all on the norm. The numerical range of an element of a normed algebra is a subset of the scalar field that reflects both the algebraic and the norm structure. We may regard it as giving information on the geometry of the normed algebra. In the case of a unital normed algebra, the numerical range of an element a is determined by the values of the given algebra-norm on the two-dimensional linear subspace spanned by a and the unit element.

In §2 we derive the elementary facts about the numerical range of an element of a unital normed algebra. We show in particular that the numerical range is a compact convex set. When the algebra is complete, the numerical range of an element contains the part of the spectrum that lies in the scalar field. We derive an important formula for the maximum of the real parts of numbers in the numerical range. We also consider the effect of computing the numerical range with respect to equivalent algebra-norms. Finally we introduce the notion of joint numerical range and indicate its relation with the joint spectrum.

In §3 we study the exponential function in a unital Banach algebra and give formulae, in terms of the exponential function, for the maximum of the real parts of numbers in the numerical range and in the spectrum of an element. This leads to the concept of a dissipative element in a unital Banach algebra.

The numerical radius of an element of a unital normed algebra is discussed in §4, where it is proved that, for complex algebras, the numerical radius is a linear-norm equivalent to the given norm. Several applications of this result are given, and some progress is made on the problem as to whether the numerical radius satisfies the power inequality. Finally we introduce the concept of the numerical index of a normed algebra.

2. ELEMENTARY THEOREMS

Notation. Let A denote a unital normed algebra over a field F, where F is either the real field R or the complex field C. Thus A is a normed algebra with a unit element 1 such that $\|1\| = 1$. Let A' denote the dual space of A, i. e. the Banach space of all continuous linear functionals on A. Let

$$S(A) = \{x \in A : \|x\| = 1\}$$

and, given $x \in S(A)$, let

$$D(A, x) = \{f \in A' : f(x) = 1 = \|f\| \} \ .$$

The Hahn-Banach theorem guarantees that $D(A, x)$ is non-empty for each $x \in S(A)$. Each element f of $D(A, x)$ is called a support functional for the unit ball at x. The elements of $D(A, 1)$ are usually called normalized states (or simply states).

Definition 1.   Given $a \in A$ and $x \in S(A)$, let

$$V(A, a, x) = \{f(ax) : f \in D(A, x)\}$$

and let

$$V(A, a) = \cup \{V(A, a, x) : x \in S(A)\}$$

$$v(a) = \sup\{|\lambda| : \lambda \in V(A, a)\}.$$

$V(A, a)$ is called the <u>numerical range</u> of a, and $v(a)$ is called the <u>numerical radius</u> of a. We may write the numerical range as $V(a)$ when it is clear to which algebra it refers.

The above definitions apply equally if **A** has no unit element, but we shall see that the presence of the unit element has useful consequences which distinguish the theory in the present chapter from the theory of the spatial numerical range of an operator as developed in the thirdchapter. Given $a, b \in A$, $\alpha, \beta \in \underset{\sim}{F}$, it is obvious that

$$V(A, a + b) \subset V(A, a) + V(A, b) \tag{1}$$

$$V(A, \alpha + \beta a) = \alpha + \beta V(A, a) \tag{2}$$

$$V(A, a, 1) = \{f(a): f \in D(A, 1)\} \tag{3}$$

and these simple facts will often be used without comment.

Lemma 2. $V(A, a) = V(A, a, 1)$ $(a \in A)$.

Proof.   Since $1 \in S(A)$ we have $V(A, a, 1) \subset V(A, a)$. Given $\lambda \in V(A, a)$, there exists $y \in S(A)$ and $g \in D(A, y)$ such that

$\lambda = g(ay)$. Let

$$f(x) = g(xy) \quad (x \in A) .$$

Then $f \in D(A, 1)$ and

$$\lambda = g(ay) = f(a) = f(a1) \in V(A, a, 1) .$$

Theorem 3. For each $a \in A$, $V(A, a)$ is a compact convex subset of $\underset{\sim}{F}$.

Proof. We have

$$D(A, 1) = \{f \in A' : \|f\| \leq 1 \text{ and } f(1) = 1\} ,$$

and so $D(A, 1)$ is a convex weak* compact subset of A'. The set $V(A, a, 1)$ is the image of $D(A, 1)$ under the weak* continuous linear mapping $f \to f(a)$, and so is a compact convex subset of $\underset{\sim}{F}$. Now apply Lemma 2.

Theorem 4. Let B be a subalgebra of A containing the unit element. Then, for each $b \in B$,

$$V(B, b) = V(A, b) .$$

Proof. By the Hahn-Banach theorem, the restriction mapping $f \to f|_B$ maps $D(A, 1)$ onto $D(B, 1)$. Therefore

$$V(A, b, 1) = V(B, b, 1)$$

and Lemma 2 completes the proof.

It follows from the above theorem that the numerical range of a is unaltered by replacing A by its completion. This allows us to assume in places that the normed algebra A is complete. Moreover V(A, a) may be computed from the subalgebra B generated by 1 and a, or from the closure of B.

The following theorem (essentially due to Lumer [40]) shows that V(A, a) is in fact determined by the values of the algebra-norm on the linear subspace spanned by 1 and a. We have $\underset{\sim}{F} = \underset{\sim}{R}$ or $\underset{\sim}{C}$.

Theorem 5. For each a ∈ A,

$$\max \{ \text{Re } \lambda : \lambda \in V(A, a) \} = \inf_{\alpha > 0} \frac{1}{\alpha} \{ \| 1 + \alpha a \| - 1 \} =$$

$$\lim_{\alpha \to 0+} \frac{1}{\alpha} \{ \| 1 + \alpha a \| - 1 \} .$$

Proof. Let $\mu = \max \{ \text{Re } \lambda : \lambda \in V(A, a) \}$. Given $\alpha > 0$, f ∈ D(A, 1), we have

$$f(a) = \frac{1}{\alpha} \{ f(1 + \alpha a) - 1 \}$$

and so

$$\text{Re } f(a) \leq \frac{1}{\alpha} \{ \| 1 + \alpha a \| - 1 \} .$$

Therefore

$$\mu \leq \inf_{\alpha > 0} \frac{1}{\alpha} \{ \| 1 + \alpha a \| - 1 \} . \tag{4}$$

We assume now that $a \neq 0$, the theorem being obvious if $a = 0$. Let $0 < \alpha < \|a\|^{-1}$. Given $x \in S(A)$, $f \in D(A, x)$, we have

$$\|(1 - \alpha a)x\| \geq \text{Re } f((1 - \alpha a)x) \geq 1 - \alpha \mu \geq 1 - \alpha \|a\| > 0 .$$

Therefore

$$\|(1 - \alpha a)x\| \geq (1 - \alpha \mu) \|x\| \quad (x \in A) .$$

Taking $x = 1 + \alpha a$, we obtain

$$\|1 + \alpha a\| \leq (1 - \alpha \mu)^{-1} \|1 - \alpha^2 a^2\|$$
$$\leq (1 - \alpha \mu)^{-1} (1 + \alpha^2 \|a^2\|) .$$

Therefore

$$\frac{\|1 + \alpha a\| - 1}{\alpha} \leq \frac{\mu + \alpha \|a^2\|}{1 - \alpha \mu} \quad (0 < \alpha < \|a\|^{-1}) .$$

Together with the inequality (4), this completes the proof.

The reader will have noticed that Theorems 3 and 4 come from the consideration of $D(A, 1)$ alone, whereas in Theorem 5 we need to use all the sets $D(A, x)$.

Recall that an element x of A is said to be <u>invertible</u> if there exists y in A such that $xy = yx = 1$.

Let $\underset{\sim}{F} = \underset{\sim}{C}$ and let A be complete. Then, for each $a \in A$, the <u>spectrum</u> of a, $Sp(A, a)$, is defined by

$$Sp(A, a) = \{\lambda \in \underset{\sim}{C} : \lambda - a \text{ is not invertible}\} .$$

18

Let $\underset{\sim}{F} = \underset{\sim}{R}$, let A be complete, and let A_C denote the complexification of A (see Rickart [57] page 6). Then, by Rickart [57] Theorem (1.3.2), $A_{\underset{\sim}{C}}$ is a complex unital Banach algebra. For each $a \in A$, the spectrum of a, $Sp(A, a)$, is now defined by

$$Sp(A, a) = Sp(A_{\underset{\sim}{C}}, a) \; .$$

The concept of the spectrum of an element can of course be extended to an arbitrary Banach algebra. (For an arbitrary normed algebra there is more than one possible definition of the spectrum of an element.) For our purposes it will be sufficient to consider the case of unital Banach algebras.

Let A be an arbitrary normed algebra. For each $a \in A$, the spectral radius of a, $\rho(a)$, is defined by

$$\rho(a) = \inf \{ \, \| a^n \|^{\frac{1}{n}} : \; n = 1, 2, 3, \dots \, \} \; .$$

It is well known (see Rickart [57] Theorem (1.4.1)) that

$$\rho(a) = \lim_{n \to \infty} \| a^n \|^{\frac{1}{n}} \; .$$

If A is a unital Banach algebra (real or complex), it is also well known (see Rickart [57] Theorem (1.6.4)) that

$$\rho(a) = \max \{ \, |\lambda| : \lambda \in Sp(A, a) \, \} \; .$$

Theorem 6. Let A be complete. Then, for each $a \in A$,

$$Sp(A, a) \cap \underset{\sim}{F} \subset V(A, a) \; .$$

Proof. Let $\lambda \in \mathrm{Sp}(A, a) \cap \underset{\sim}{F}$. Then $\lambda - a$ is a singular element of A. Suppose that it has no left inverse, and let $J = A(\lambda - a)$. Then J is a proper left ideal of A. Since A is a unital Banach algebra,

$$\| 1 - x \| \geq 1 \quad (x \in J) \ .$$

By the Hahn-Banach theorem, there exists $f \in A'$ such that $f(1) = \| f \| = 1$ and $f(J) = \{0\}$. Thus $f \in D(A, 1)$ and $f(\lambda - a) = 0$. This gives $\lambda = f(a) \in V(A, a, 1)$. A similar proof in terms of right ideals is available if $\lambda - a$ has no right inverse.

Remark. We recall that the mapping $a \to \mathrm{Sp}(A, a)$ is upper semi-continuous (see Rickart [57] Theorem (1.6.16)). It is easy to verify that the mapping $a \to V(A, a)$ is also upper semi-continuous.

Notation. An algebra-norm p on A is said to be equivalent to the given norm if there exist $\alpha, \beta > 0$ such that

$$\alpha \| a \| \leq p(a) \leq \beta \| a \| \quad (a \in A) \ .$$

Let $\underline{N}$ denote the set of all algebra-norms p on A that are equivalent to the given norm and satisfy $p(1) = 1$. Given $p \in \underline{N}$, let $V_p(A, a)$ denote the numerical range of a determined by the norm p in place of the given norm. Given $E \subset \underset{\sim}{F}$, let $\mathrm{co}\ E$ denote the convex hull of E, i.e. the least convex set containing E. We need next two results of independent interest.

Lemma 7. Let S be a bounded multiplicative semi-group in A. Then there exists $p \in \underline{N}$ such that $p(s) \le 1$ $(s \in S)$.

Proof. There is no loss of generality in assuming that $1 \in S$. First take

$$q(x) = \sup \{ \| sx \| : s \in S \} \quad (x \in A) .$$

Let $M = \sup \{ \| s \| : s \in S \}$. It is easily verified that q is an algebra-norm on A such that

$$\| x \| \le q(x) \le M \| x \| \quad (x \in A)$$

and

$$q(sx) \le q(x) \quad (s \in S, \ x \in A) .$$

Now take

$$p(a) = \sup \{ q(ax) : x \in A, q(x) \le 1 \} \quad (a \in A)$$

and verify that $p \in \underline{N}$, $p(s) \le 1$ $(s \in S)$.

Lemma 8. Let $a_1, \ldots, a_n$ be mutually commuting elements of A and let $\epsilon > 0$. Then there exists $p \in \underline{N}$ such that

$$p(a_k) < \rho(a_k) + \epsilon \quad (k = 1, \ldots, n) .$$

Proof. Let $b_k = a_k/(\rho(a_k) + \epsilon)$ $(k = 1, \ldots, n)$, and let S be the multiplicative semigroup in A generated by $b_1, \ldots, b_n$. Since $b_1, \ldots, b_n$ are mutually commuting elements and $\rho(b_k) < 1$ for each k, it follows that S is a bounded semigroup in A. Lemma 7 gives $p \in \underline{N}$ with

$$p(b_k) \leq 1 \qquad (k = 1, \ldots, n) ,$$

and therefore

$$p(a_k) \leq \rho(a_k) + \epsilon \qquad (k = 1, \ldots, n) .$$

Corollary 9. Given $a \in A$,

$$\rho(a) = \inf \{p(a) : p \in \underline{N}\} .$$

Proof. The above lemma shows that

$$\inf \{p(a) : p \in \underline{N}\} \leq \rho(a) .$$

Since, for each $p \in \underline{N}$,

$$\rho(a) = \lim_{n \to \infty} p(a^n)^{\frac{1}{n}} = \inf_n p(a^n)^{\frac{1}{n}} \leq p(a) ,$$

the proof is complete.

We remark that Lemma 8 also gives an easy proof of the inequalities $\rho(a + b) \leq \rho(a) + \rho(b)$, $\rho(ab) \leq \rho(a)\rho(b)$ for commuting elements a, b.

Theorem 10. Let A be complete and let $\underset{\sim}{F} = \underset{\sim}{C}$. Then, for each $a \in A$,

$$\text{co } Sp(A, a) = \cap \{ V_p(A, a) : p \in \underline{N} \} .$$

Proof. It is immediate from Theorems 3 and 6 that

$$\text{co } Sp(A, a) \subset \cap \{ V_p(A, a) : p \in \underline{N} \}.$$

Since $Sp(A, a)$ is compact, $\text{co } Sp(A, a)$ is a compact convex set and is therefore the intersection of the open circular discs containing $Sp(A, a)$. Suppose then that

$$|\lambda - \alpha| < r \quad (\lambda \in Sp(A, a)) .$$

Then $\rho(a - \alpha) < r$ and so, by Lemma 8, there is $p \in \underline{N}$ with $p(a - \alpha) < r$. But then it is obvious that

$$|\lambda - \alpha| < r \quad (\lambda \in V_p(A, a)) ,$$

and so $\cap \{ V_p(A, a) : p \in \underline{N} \}$ is contained in every open circular disc that contains $Sp(A, a)$. The proof is complete.

Definition 11. Let A be a unital normed algebra and let $a_1, \ldots, a_n \in A$. The joint numerical range of $a_1, \ldots, a_n$, $V(A; a_1, \ldots, a_n)$, is defined by

$$V(A; a_1, \ldots, a_n) = \{ (f(a_1), \ldots, f(a_n)) : f \in D(A, 1) \} .$$

It is clear that $V(A; a_1, \ldots, a_n)$ is a compact convex subset of $\underset{\sim}{F}^n$.

Let $\underset{\sim}{F} = \underset{\sim}{C}$, let A be complete and let $a_1, \ldots, a_n \in A$. The $\underline{\text{joint spectrum}}$ of $a_1, \ldots, a_n$, $Sp(A; a_1, \ldots, a_n)$, is the set of complex n-tuples $(\lambda_1, \ldots, \lambda_n)$ such that either

$$A(\lambda_1 - a_1) + \ldots + A(\lambda_n - a_n)$$

is a proper left ideal or

$$(\lambda_1 - a_1)A + \ldots + (\lambda_n - a_n)A$$

is a proper right ideal. For a real Banach algebra A, the $\underline{\text{joint}}$ $\underline{\text{spectrum}}$ of $a_1, \ldots, a_n$, $Sp(A; a_1, \ldots, a_n)$, is defined to be the joint spectrum of $a_1, \ldots, a_n$ in the complexification of A.

Theorem 12. Let A be complete and let $a_1, \ldots, a_n \in A$. Then
$$Sp(A; a_1, \ldots, a_n) \cap \underset{\sim}{F} \subset V(A; a_1, \ldots, a_n) .$$

Proof. Similar to Theorem 6.

Given $p \in \underline{N}$, let $V_p(A; a_1, \ldots, a_n)$ denote the joint numerical range of $a_1, \ldots, a_n$ determined by the norm p in place of the given norm.

Theorem 13. Let $\underset{\sim}{F} = \underset{\sim}{C}$, let A be complete and let $a_1, \ldots, a_n$ be mutually commuting elements of A. Then

$$co\, Sp(A; a_1, \ldots, a_n) = \cap \{V_p(A; a_1, \ldots, a_n): p \in \underline{N}\} .$$

Proof. It is immediate from Theorem 12 and the convexity of the joint numerical range that

$$\text{co Sp}(A; a_1, \ldots, a_n) \subset \cap \{V_p(A; a_1, \ldots, a_n): p \in \underline{N}\}.$$

An $n \times n$ matrix T with complex entries may be regarded as a linear mapping of A^n into A^n as well as of $\underset{\sim}{C}^n$ into $\underset{\sim}{C}^n$. Given $\underset{\sim}{a} = (a_1, a_2, \ldots, a_n) \in A^n$, it is straightforward to verify that $V(A; T\underset{\sim}{a}) = TV(A; \underset{\sim}{a})$, and therefore that

$$\cap \{V_p(A; T\underset{\sim}{a}): p \in \underline{N}\} \supset T[\cap\{V_p(A; \underset{\sim}{a}): p \in \underline{N}\}] .$$

If T is non-singular, the opposite inclusion follows and we have equality; also, by a similar argument,

$$\text{Sp}(A; T\underset{\sim}{a}) = T \text{ Sp}(A; \underset{\sim}{a}) .$$

Suppose $\underset{\sim}{\alpha} = (\alpha_1, \ldots, \alpha_n) \in \underset{\sim}{C}^n \setminus \text{co Sp}(A; \underset{\sim}{a})$. Then there exists a linear functional ϕ on $\underset{\sim}{C}^n$ and a real number r such that

$$\text{Re } \phi(\underset{\sim}{\zeta}) < r < \text{Re } \phi(\alpha) \quad (\underset{\sim}{\zeta} \in \text{co Sp}(A; \underset{\sim}{a})) .$$

Let $\phi(\underset{\sim}{\zeta}) = t_{11}\zeta_1 + \ldots + t_{1n}\zeta_n$ $(\underset{\sim}{\zeta} = (\zeta_1, \ldots, \zeta_n) \in C^n)$, and choose a non-singular $n \times n$ matrix T with $(t_{11}, \ldots, t_{1n})$ as its first row. Then

$$\text{Re } \zeta_1 < r < \text{Re } \beta_1 \quad (\underset{\sim}{\zeta} = (\zeta_1, \ldots, \zeta_n) \in \text{Sp}(A; T\underset{\sim}{a})) ,$$

where $(\beta_1, \ldots, \beta_n) = T\underset{\sim}{\alpha}$. It follows that there exists an open polydisc containing $\text{Sp}(A; T\underset{\sim}{a})$ that excludes $T\underset{\sim}{\alpha}$. Since the elements $a_1, \ldots, a_n$ commute with each other, the proof is completed as in Theorem 10 by an application of Lemma 8.

Remark.   We owe the proof of Theorem 13 to M. J. Crabb (unpublished).   Most of the other material in this section may be found in Bonsall [11].

3. THE EXPONENTIAL FUNCTION

The exponential function in a unital Banach algebra has many fruitful applications. We derive some of its properties in this section using elementary methods.

Throughout this section A denotes a unital Banach algebra over $\underset{\sim}{F}$. Let G(A) denote the set of invertible elements of A.

Definition 1.   Let $a \in A$. Then exp(a) is defined by

$$\exp(a) = 1 + \sum_{n=1}^{\infty} \frac{1}{n!} a^n .$$

Theorem 2.   Let $a, b \in A$ and ab = ba.   Then

(i)   exp(a + b) = exp(a) exp(b)

(ii)   exp(a) exp(-a) = 1

(iii)   exp(a) $\in$ G(A) .

Proof.   Let $x_n, y_n, z_n, \xi_n, \eta_n, \zeta_n$ be defined by

$$x_n = 1 + \sum_{k=1}^{n} \frac{1}{k!} a^k, \quad y_n = 1 + \sum_{k=1}^{n} \frac{1}{k!} b^k ,$$

$$z_n = 1 + \sum_{k=1}^{n} \frac{1}{k!} (a + b)^k , \quad \xi_n = 1 + \sum_{k=1}^{n} \frac{1}{k!} \|a\|^k ,$$

26

$$\eta_n = 1 + \sum_{k=1}^{n} \frac{1}{k!} \, \|b\|^k \,, \quad \zeta_n = 1 + \sum_{k=1}^{n} \frac{1}{k!} \, (\|a\| + \|b\|)^k \,.$$

We have

$$x_n y_n - z_n = \sum_{j,\,k=1}^{n} \alpha_{jk} \, a^j b^k$$

with $\alpha_{jk} \geq 0$ for all j, k. Therefore

$$\|x_n y_n - z_n\| \leq \sum_{j,\,k=1}^{n} \alpha_{jk} \|a\|^j \|b\|^k = \xi_n \eta_n - \zeta_n \,.$$

But $\xi_n \eta_n - \zeta_n \to \exp(\|a\|) \exp(\|b\|) - \exp(\|a\| + \|b\|) = 0$ as $n \to \infty$. This proves (i), and (ii), (iii) are then clear.

Part (iii) asserts that the range of the exponential function is a subset of $G(A)$. In fact if A is commutative then $\exp(A) = G_1(A)$, where $G_1(A)$ is the principal component of $G(A)$, i. e. the maximal connected subset of $G(A)$ that contains 1. For an arbitrary Banach algebra, $G_1(A)$ is the subgroup of $G(A)$ generated by $\exp(A)$ (see Rickart [57] Theorem (1.4.10)). This result, for the commutative case, will be used in §6.

Theorem 3. Let $a \in A$. Then $\exp(a) = \lim_{n \to \infty} (1 + \frac{1}{n} a)^n$.

Proof. Let x_n, y_n, ξ_n, η_n be defined by

$$x_n = 1 + \sum_{k=1}^{n} \frac{1}{k!} a^k \,, \quad y_n = (1 + \frac{1}{n} a)^n \,,$$

$$\xi_n = 1 + \sum_{k=1}^{n} \frac{1}{k!} \, \|a\|^k \,, \quad \eta_n = (1 + \frac{1}{n} \|a\|)^n \,.$$

We have

$$y_n = 1 + \frac{1}{1!}\, a + (1 - \frac{1}{n})\frac{1}{2!}\, a^2 + \ldots + (1 - \frac{1}{n})(1 - \frac{2}{n}) \ldots$$

$$\ldots (1 - \frac{n-1}{n})\frac{1}{n!}\, a^n \, ,$$

and so

$$x_n - y_n = \sum_{k=2}^{n} \alpha_k a^k$$

with $\alpha_k \geq 0$ $(k = 2, \ldots, n)$. Therefore

$$\|x_n - y_n\| \leq \sum_{k=2}^{n} \alpha_k \|a\|^k = \xi_n - \eta_n \, .$$

But $\lim\limits_{n \to \infty} (\xi_n - \eta_n) = 0$ since $\exp(\|a\|) = \lim\limits_{n \to \infty} (1 + \frac{1}{n}\|a\|)^n$.

Theorem 4. Let $a \in A$. Then

$$\max \{\operatorname{Re} \lambda : \lambda \in V(A, a)\} = \sup \{\frac{1}{\alpha} \log \|\exp(\alpha a)\| : \alpha > 0\}$$

$$= \lim_{\alpha \to 0+} \frac{1}{\alpha} \log \|\exp(\alpha a)\| \, .$$

Proof. Let $\mu = \max \operatorname{Re} V(A, a)$, and $\alpha \geq 0$. Given $x \in S(A)$, $f \in D(A, x)$, we have, as in the proof of Theorem 2.5,

$$\|(1 - \alpha a)x\| \geq (1 - \alpha\mu)\|x\| \qquad (x \in A) \, .$$

If $1 - \alpha\mu \geq 0$, we have by induction

$$\|(1 - \alpha a)^n x\| \geq (1 - \alpha\mu)^n \|x\| \qquad (x \in A, \ n = 1, 2, \ldots) \, . \quad (1)$$

We have $1 - \dfrac{\alpha}{n}\mu \geq 0$ for all sufficiently large n. Therefore, replacing α by $\dfrac{\alpha}{n}$ in (1) and letting $n \to \infty$, we obtain

$$\| \exp(-\alpha a)x \| \geq \exp(-\alpha\mu) \| x \| \ .$$

Taking $x = \exp(\alpha a)$, we get

$$\| \exp(\alpha a) \| \leq \exp(\alpha\mu) \ ,$$

and so

$$\sup \{ \tfrac{1}{\alpha} \log \| \exp(\alpha a) \| : \alpha > 0 \} \leq \mu \ . \tag{2}$$

On the other hand, we have

$$\| \exp(\alpha a) \| = \| 1 + \alpha a \| + \lambda(\alpha) \ ,$$

where, for some $M > 0$,

$$|\lambda(\alpha)| \leq M\alpha^2 \quad (0 \leq \alpha \leq 1) \ . \tag{3}$$

Using the inequality

$$\log t \geq \frac{t-1}{t} \quad (t > 0) \ ,$$

we have, for $\alpha > 0$,

$$\tfrac{1}{\alpha} \log \| \exp(\alpha a) \| \geq \frac{\frac{1}{\alpha}\{ \| 1 + \alpha a \| - 1 \} + \frac{1}{\alpha} \lambda(\alpha)}{\| 1 + \alpha a \| + \lambda(\alpha)} \ .$$

Using (3) and Theorem 2. 5 we see that the right hand side of the inequality converges to μ as $\alpha \to 0+$. Combined with (2), this completes the proof.

The definition below is essentially due to Lumer and Phillips [43]; the original motivation lay in initial value problems in the theory of partial differential equations.

Definition 5. An element a of A is said to be <u>dissipative</u> if

$$\text{Re } \lambda \leq 0 \quad (\lambda \in V(A, a)) .$$

Theorem 6. (Lumer-Phillips). Let $a \in A$. Then a is dissipative if and only if $\|\exp(ta)\| \leq 1$ $(t \geq 0)$.

Proof. Apply Theorem 4.

Remark. If the semi-group $\{\exp(ta): t \geq 0\}$ is bounded, Lemma 2.7 gives an equivalent algebra-norm p on A such that $p(\exp(ta)) \leq 1$ $(t \geq 0)$ and hence a is dissipative with respect to the norm p.

A conjecture of Bohnenblust and Karlin [10] is equivalent to the following question. Is 0 the only quasi-nilpotent dissipative element of a Banach algebra? Lumer and Phillips showed that the answer was negative ([43] Theorem 2.2, and Theorem 9.4). On the other hand they showed that if a is a quasi-nilpotent dissipative element in a complex unital Banach algebra such that, for some $k > 0$, $\exp(\alpha a) = O(\alpha^k)$ as $\alpha \to -\infty$, then $a = 0$. They also obtained results for elements whose numerical range is contained within an angle, and these results were sharpened by Stampfli and Williams [67].

The theorem below is the analogue of Theorem 4 for the spectrum of an element. We need first a technical lemma.

Lemma 7. Let f be a continuous subadditive mapping of the positive reals into $\underset{\sim}{R}$. Then

$$\lim_{t \to +\infty} \frac{1}{t} f(t) = \inf \{ \frac{1}{t} f(t): t > 0 \} .$$

Proof. Given $\alpha > \inf \{ \frac{1}{t} f(t): t > 0 \}$, choose $s > 0$ such that $\frac{1}{s} f(s) < \alpha$. Since f is continuous,

$$\sup \{ f(t): s \le t \le 2s \} = m < \infty .$$

Therefore, for all positive integers n and real t such that

$$(n + 1)s \le t \le (n + 2)s , \tag{4}$$

we have

$$f(t) \le f(ns) + f(t - ns) \le nf(s) + m ,$$

and so

$$\frac{1}{t} f(t) < \frac{ns}{t} \alpha + \frac{m}{t} .$$

Letting $t \to \infty$, and $n \to \infty$ such that (4) holds, we conclude that

$$\limsup_{t \to +\infty} \frac{1}{t} f(t) \le \alpha ,$$

from which the result follows.

Theorem 8. Let $a \in A$. Then

$$\max \{ \operatorname{Re} \lambda : \lambda \in \operatorname{Sp}(A, a) \} = \inf \{ \frac{1}{\alpha} \log \| \exp(\alpha a) \| : \alpha > 0 \}$$

$$= \lim_{\alpha \to +\infty} \frac{1}{\alpha} \log \| \exp(\alpha a) \| .$$

Proof. We have $\rho(\exp(a)) > 0$, and so

$$\log \rho(\exp(a)) = \lim_{n \to \infty} \frac{1}{n} \log \| \exp na \|$$

$$= \inf \{ \frac{1}{\alpha} \log \| \exp(\alpha a) \| : \alpha > 0 \} ,$$

by Lemma 7. The spectral mapping theorem gives $\operatorname{Sp}(A, \exp(a)) = \exp \operatorname{Sp}(A, a)$, and so

$$\rho(\exp(a)) = \max \{ \exp \operatorname{Re} \lambda : \lambda \in \operatorname{Sp}(A, a) \}$$

$$= \exp(\max \operatorname{Re} \operatorname{Sp}(A, a)) .$$

This gives the result.

Corollary 9. $\rho(a) = \max_{|z|=1} \lim_{\alpha \to +\infty} \frac{1}{\alpha} \log \| \exp(\alpha z a) \| .$

The relevance of the following theorem to the numerical range is not immediately apparent, but we need it later and it is convenient to insert it here since its proof needs the exponential function.

Theorem 10. (Le Page [39], Hirschfeld and Zelazko [32]).
Let $\underset{\sim}{F} = \underset{\sim}{C}$. If, for some $k > 0$,

$$\rho(a) \geq k\|a\| \qquad (a \in A) ,$$

then **A** is commutative.

Proof. Let $a, b \in A$, and let

$$\phi(\lambda) = \exp(-\lambda a) \; b \; \exp(\lambda a) \qquad (\lambda \in \underset{\sim}{C}) .$$

Since $Sp(A, \phi(\lambda)) = Sp(A, b)$, we have

$$k\|\phi(\lambda)\| \leq \rho(\phi(\lambda)) \leq \rho(b) \qquad (\lambda \in \underset{\sim}{C}) .$$

Let $f \in A'$ with $\|f\| \leq 1$. Then $f \circ \phi$ is a bounded entire function and so is constant by Liouville's theorem. Therefore, by the Hahn-Banach theorem, ϕ is constant and so $\phi(1) = \phi(0)$. This gives

$$\exp(a) \; b = b \; \exp(a) . \qquad (5)$$

Replace a by λa in (5) with $\lambda \in \underset{\sim}{C}$, expand in powers of λ, and equate coefficients of λ to obtain $ab = ba$.

4. NUMERICAL RADIUS THEOREMS

Let **A** be a unital normed algebra and let $a \in A$. Recall that the numerical radius of a is given by

$$v(a) = \max\{\,|\lambda| : \lambda \in V(A, a)\,\} .$$

It is clear from formulae (1) and (2) of §2 that v is a linear semi-norm on **A**. The result below shows that if $\underset{\sim}{F} = \underset{\sim}{C}$ then v is a linear-norm on A that is equivalent to the given norm. This crucial result appears slightly hidden in Bohnenblust and Karlin [10] (Theorem 1 together with the inequality $\|x\| \leq e \, \Psi(x)$ which occurs on page 219). A simpler proof was given by Lumer [40], though with the constant $\frac{1}{4}$ in place of $\frac{1}{e}$. The present proof borrows ideas from Lumer and from Glickfeld [28].

Theorem 1. Let $\underset{\sim}{F} = \underset{\sim}{C}$. Then, for all $a \in A$,

$$\|a\| \geq v(a) \geq \frac{1}{e} \|a\| .$$

Proof. The first inequality is obvious. We prove the second. By Theorem 2.4 we may suppose that A is complete, for replacement of A by its completion does not change $V(A, a)$. Let $b \in A$ and

$$v(b) \leq \mu < 1 . \tag{1}$$

Given $x \in S(A)$, there exists $f \in D(A, x)$, and so for all complex numbers λ with $|\lambda| \leq 1$ we have

$$\|(1 - \lambda b)x\| \geq |f((1 - \lambda b)x| = |1 - \lambda f(bx)| \geq 1 - \mu .$$

Therefore

$$\|(1 - \lambda b)x\| \geq (1 - \mu)\|x\| \qquad (x \in A, \ |\lambda| \leq 1) . \tag{2}$$

By Theorem 2.6, $\rho(b) \leq \mu < 1$, and so $1 - \lambda b$ is invertible whenever $|\lambda| \leq 1$. It follows from (2) that

$$\left\| (1 - \lambda b)^{-1} \right\| \le (1 - \mu)^{-1} \quad (|\lambda| \le 1) . \tag{3}$$

With $\omega_1, \ldots, \omega_n$ denoting the n-th roots of unity, we have for all integers j

$$\sum_{k=1}^{n} \omega_k^{j} = \begin{cases} 0 & (j \not\equiv 0 \pmod{n}) \\ n & (j \equiv 0 \pmod{n}) . \end{cases} \tag{4}$$

Consider $S(r, n)$ given for $r = 1, 2, \ldots,$ by

$$S(r, n) = \frac{1}{n} \sum_{k=1}^{n} \omega_k^{-1} (1 - \omega_k b)^{-r} .$$

We have

$$\omega_k^{-1} (1 - \omega_k b)^{-r} = \omega_k^{-1} \{ 1 + r\omega_k b + \frac{r(r+1)}{2!} (\omega_k b)^2 + \ldots \} ,$$

and therefore

$$S(r, n) = rb + \frac{r(r+1)\ldots(r+n)}{(n+1)!} b^{n+1} + \ldots .$$

Since $\rho(b) < 1,$ it follows that

$$\lim_{n \to \infty} S(r, n) = rb .$$

By (3), $\| S(r, n) \| \le (1 - \mu)^{-r},$ and therefore

$$r \|b\| \le (1 - \mu)^{-r} \quad (r = 1, 2, \ldots) . \tag{5}$$

Now let a be an arbitrary element of A, let $\kappa > v(a),$ $r \ge 2,$ $\mu = \frac{1}{r},$ and take $b = \frac{1}{r\kappa} a.$ Then b, μ satisfy (1) and therefore, by (5),

$$\frac{1}{\kappa} \|a\| \le (1 - \frac{1}{r})^{-r} \quad (r = 2, 3, \ldots) \ ,$$

from which $\kappa \ge \frac{1}{e} \|a\|$. This completes the proof.

The corresponding statement for real algebras is false. To see this, let $A = \underset{\sim}{C}$ as a real Banach algebra with modulus as norm, and let $a = i$. If $f \in D(A, 1)$, then, for some real q,

$$f(x + iy) = x + qy \quad (x, y \in \underset{\sim}{R}) \ .$$

Since $\|f\| = 1$, we have

$$1 + qt \le (1 + t^2)^{\frac{1}{2}} \quad (t \in \underset{\sim}{R}) \ .$$

This gives $q = 0$, and therefore $V(A, a) = \{0\}$, $v(a) = 0$. It is, however, proved in Bonsall and Duncan [14] that if A is a real unital Banach algebra then $a = 0$ whenever $V(A, a) = V(A, a^2) = \{0\}$. The proof in [14] is elementary, and we shall see later that a stronger result (Theorem 5.15) can be proved by using deeper methods.

We give next some of the applications of Theorem 1.

Definition 2. A point $u \in S(A)$ is called a <u>vertex</u> of the unit ball if $D(A, u)$ is a total subset of A', i.e. if $a = 0$ whenever $f(a) = 0$ $(f \in D(A, u))$. It is easily seen that every vertex of the unit ball is an extreme point of the unit ball (see the remark following the proof of Theorem 6).

Definition 3. A point $u \in S(A)$ is called a point of <u>local uniform convexity</u> for the unit ball if

$$x_n, \; y_n \; \epsilon \; S(A), \quad \lim_{n \to \infty} \; (x_n + y_n) = 2u$$

together imply that

$$\lim_{n \to \infty} \; x_n = \lim_{n \to \infty} \; y_n = u \, .$$

In the following theorem the first part is due to Bohnenblust and Karlin [10] and the second to Lumer [40].

We need first a simple lemma.

Lemma 4. Let $\alpha, \beta \; \epsilon \; \underset{\sim}{C}$ with $|\alpha| \leq 1$, $|\beta| \leq 1$, and let $\lambda = \frac{1}{2} \, (\alpha + \beta)$. Then

$$\max(|1 - \alpha|, \; |1 - \beta|) \leq 2|1 - \lambda|^{\frac{1}{2}}$$

whenever $|1 - \lambda| \leq 1$.

Proof. Let $a = |1 - \alpha|$, $b = |1 - \beta|$, $c = |1 - \lambda|$, $d = \frac{1}{2} \, |\alpha - \beta|$. The parallelogram law gives

$$2|\lambda|^2 + 2d^2 = |\alpha|^2 + |\beta|^2 \leq 2 \, .$$

Since $|\lambda| \geq 1 - c \geq 0$, we have

$$d^2 + (1 - c)^2 \leq d^2 + |\lambda|^2 \leq 1 \, ,$$

and therefore $d^2 + c^2 \leq 2c$. Using the parallelogram law again we obtain

$$a^2 + b^2 = 2d^2 + 2c^2 \leq 4c \, .$$

The result follows.

> **Theorem** 5. Let $\underset{\sim}{F} = \underset{\sim}{C}$. Then the unit element of A is a vertex and a point of local uniform convexity of the unit ball.

> **Proof.** Let $a \in A$, and suppose that $f(a) = 0$ for all $f \in D(A, 1)$. Then $V(A, a) = \{0\}$, and so $a = 0$ by Theorem 1. This shows that 1 is a vertex of the unit ball.
> Let $a, b \in S(A)$ with $v(1 - \frac{1}{2}(a + b)) \leq 1$. Using Lemma 4, we then obtain

$$v(1 - a) = \max\{|1 - f(a)| : f \in D(A, 1)\}$$

$$\leq \max\{2|1 - f(\tfrac{1}{2}(a + b))|^{\frac{1}{2}} : f \in D(A, 1)\}$$

$$= 2\, v(1 - \tfrac{1}{2}(a + b))^{\frac{1}{2}}. \tag{6}$$

Now let $x_n, y_n \in S(A)$ with $\lim_{n \to \infty} (1 - \frac{1}{2}(x_n + y_n)) = 0$. Then $\lim_{n \to \infty} v(1 - \frac{1}{2}(x_n + y_n)) = 0$, and it follows from (6) that $\lim_{n \to \infty} v(1 - x_n) = 0$. Theorem 1 now gives $\lim_{n \to \infty} x_n = 1$, and so also $\lim_{n \to \infty} y_n = 1$. The proof is complete.

Remark. In the above theorem the unit element may be replaced by any element $u \in G(A)$ such that $u, u^{-1} \in S(A)$. For then the mapping $x \to u^{-1}x$ is a linear isometry of A onto itself which maps u to 1.

Theorem 6. (Stampfli [66].) Let $F = C$, and let a be an element of A such that $\rho(a) < 2$ and $1 + \zeta a$ is invertible and satisfies $\|(1 + \zeta a)^{-1}\| \leq 1$ for all complex numbers ζ with $|\zeta| = 1$. Then $a = 0$.

Proof. The conditions are unchanged if we replace A by its completion, and so we may assume that A is complete. We show first that $\rho(a) = 0$. If not, then there exists λ with $0 < |\lambda| < 2$ and $\lambda - a$ singular. We have $\frac{1}{\lambda} = r\zeta$ with $r > \frac{1}{2}$ and $|\zeta| = 1$, and

$$1 - \frac{1}{\lambda} a = 1 - r\zeta a = r(1 - \zeta a) + 1 - r$$

$$= r(1 - \zeta a)(1 + \frac{1 - r}{r}(1 - \zeta a)^{-1}).$$

Then $\left| \frac{1 - r}{r} \right| < 1$, and so

$$\| \frac{1 - r}{r}(1 - \zeta a)^{-1} \| < 1 ,$$

which implies that $1 - \frac{1}{\lambda} a$ is invertible. This contradiction implies that $\rho(a) = 0$.

Given a positive integer n, let $\omega_2, \ldots, \omega_n$ denote the n-th roots of unity distinct from 1. Then

$$(1 - a)(1 - a^n)^{-1} = (1 - \omega_2 a)^{-1}(1 - \omega_3 a)^{-1} \ldots (1 - \omega_n a)^{-1} .$$

Therefore $\|(1 - a)(1 - a^n)^{-1}\| \leq 1$. Since $\rho(a) = 0$, $a^n \to 0$ as $n \to \infty$, and so $\|1 - a\| \leq 1$. Replacing a by $-a$, we have also $\|1 + a\| \leq 1$. Given $f \in D(A, 1)$, we therefore have

$$\left|1 \pm f(a)\right| = \left|f(1 \pm a)\right| \le \left\|1 \pm a\right\| \le 1 ,$$

and so $f(a) = 0$. Therefore, by Theorem 5, $a = 0$.

Remark. The last step in the proof is actually showing that a vertex of the unit ball is an extreme point.

We have already observed that

$$\rho(a) \le v(a) \le \|a\| \qquad (a \in A) .$$

If $\rho(a) = \|a\|$ $(a \in A)$, then, by Theorem 3.10, A is commutative and is thus isometrically isomorphic to a function algebra with the supremum norm.

Theorem 7. Let $\underset{\sim}{F} = \underset{\sim}{C}$, and let

$$\rho(a) = v(a) \qquad (a \in A) .$$

Then A is commutative and ρ is an algebra-norm equivalent to the given norm.

Proof. By Theorem 1, $\rho(a) \ge \dfrac{1}{e} \|a\|$ $(a \in A)$. Theorem 3.10 now shows that A is commutative. It follows that ρ is a Banach algebra norm on A, as required.

It is clear that the condition $\rho(a) = v(a)$ $(a \in A)$ is equivalent to the condition

$$V(a) = \text{co } Sp(A, a) \qquad (a \in A) .$$

Moreover, if for some $k > 0$, $\rho(a) \geq k\, v(a)$ $(a \in A)$, then we again conclude that A is commutative.

For an arbitrary normed algebra the spectral radius is not even a linear semi-norm, but if the algebra is commutative then the spectral radius is an algebra semi-norm. Even for commutative algebras the numerical radius can fail to be an algebra semi-norm (see Theorem 9.4 together with Halmos [30] page 116). Although the numerical radius fails to be submultiplicative, for complex B*-algebras A the power inequality

$$v(a^n) \leq v(a)^n \qquad (a \in A, \quad n = 1, 2, \ldots) \tag{7}$$

holds true (see Theorem 9.4 together with Halmos [30] page 117). Various generalizations of (7) for complex B*-algebras have been obtained by Kato [38], Berger and Stampfli [5], and Nagy and Foiaş [46]. It is not known[†] whether (7) holds for an arbitrary complex normed algebra. Theorem 1 gives the very weak inequality $v(a^n) \leq e^n v(a)^n$. The result below constitutes a considerable sharpening of the last inequality and implies in particular that

$$v(a^n) \leq e\, n^{\frac{1}{2}} v(a)^n \qquad (n = 1, 2, \ldots). \tag{8}$$

Theorem 8. (Crabb [17]). Let $F = C$. Then, for all $a \in A$,

$$\|a^n\| \leq n! \left(\frac{e}{n}\right)^n v(a)^n \qquad (n = 1, 2, \ldots).$$

---

† See Remark (10) of §14.

Proof. We may suppose that A is complete. Let $b \in A$ with $v(b) \leq \mu < 1$. Then, as in Theorem 1, $1 - \lambda b$ is invertible for complex numbers λ with $|\lambda| \leq 1$. Let m, n, r be positive integers, let $\omega_1, \ldots, \omega_m$ be the m-th roots of unity, and let $T(m, r, n)$ be given by

$$T(m, r, n) = \frac{1}{m} \sum_{k=1}^{m} \omega_k^n (1 - \omega_k^{-1} b)^{-r} \ .$$

Using formula (4), we have for $m > n$ that

$$T(m, r, n) = \frac{1}{m} \sum_{k=1}^{m} \omega_k^n (1 + r\omega_k^{-1} b + \ldots +$$

$$\frac{r(r + 1) \ldots (r + n - 1)}{1 . 2 \ldots n} \omega_k^{-n} b^n + \ldots) = \frac{r(r + 1) \ldots (r + n - 1)}{1 . 2 \ldots n} b^n +$$

$$\frac{r(r + 1) \ldots (r + n + m - 1)}{1 . 2 \ldots (n + m)} b^{n+m} + \ldots \ .$$

Since $\rho(b) < 1$, it follows that

$$\lim_{m \to \infty} T(m, r, n) = \frac{r(r + 1) \ldots (r + n - 1)}{1 . 2 \ldots n} b^n \ .$$

Using formula (3) we have $\| T(m, r, n) \| \leq (1 - \mu)^{-r}$ and therefore

$$\frac{r(r + 1) \ldots (r + n - 1)}{1 . 2 \ldots n} \| b^n \| \leq (1 - \mu)^{-r} \quad (r = 1, 2 \ldots) \ . \quad (9)$$

Now let a be an arbitrary element of A. The case $v(a) = 0$ is covered in Theorem 1 and so we may assume $v(a) > 0$. Let $b = \frac{n}{r + 1} \frac{a}{v(a)}$ and $\mu = v(b)$. Then $\mu < 1$ for $r > n - 1$, and therefore, by (9),

$$\|a^n\| \leq \frac{1.2...n}{r(r+1)...(r+n-1)} \left(\frac{r+1}{n}\right)^n \left(\frac{r-n+1}{r+1}\right)^{-r} v(a)^n .$$

Let $r \to \infty$ and the result follows.

Further sharpening of formula (8) is possible in special cases (see Theorem 10. 15).

Definition 9. Let A be a unital normed algebra. The numerical index of A is defined by

$$n(A) = \inf \{v(a): a \in A, \|a\| = 1\} .$$

In the case $\underset{\sim}{F} = \underset{\sim}{C}$, it follows from Theorem 1 that $\frac{1}{e} \leq n(A) \leq 1$. It is clear that the numerical index of A remains unaltered under an isometric isomorphism. In a lecture at the North British Functional Analysis Seminar at Edinburgh in 1968, G. Lumer suggested the possibility of characterizing complex Banach algebras by some family of such invariants.

By way of illustration let A be a complex unital B*-algebra, so that $\|a^*a\| = \|a\|^2$ $(a \in A)$. It is shown in Example 5. 3 that self-adjoint elements of A have real numerical range. Given $a \in A$ we may write $a = h + ik$ with $h^* = h$, $k^* = k$. Then

$$V(a) = \{f(h) + if(k): \; f \in D(A, 1)\},$$

and, since $V(h)$, $V(k)$ are real, it follows that $v(h) \leq v(a)$, $v(k) \leq v(a)$. For any self-adjoint element b of A, we have $\rho(b) = \|b\|$ and therefore $v(b) = \|b\|$. It follows that

$$\|a\| \leq \|h\| + \|k\| = v(h) + v(k) \leq 2\, v(a)$$

and therefore $n(A) \geq \frac{1}{2}$. The constant $\frac{1}{2}$ is best possible in general (see Theorem 9. 4 and Halmos [30] page 114).

If A is also commutative, then for each $a \in A$, $\rho(a) = \|a\|$, and therefore $n(A) = 1$. Suppose now that B is any complex unital Banach algebra with $n(B) < \frac{1}{2}$. Then no involution on B can make B into a B*-algebra. Likewise if B is commutative and $n(B) < 1$ it cannot be made into a B*-algebra.

Further results on numerical index are discussed in §9.

2. Hermitian Elements of a Complex Unital Banach Algebra

The elements of a complex unital normed algebra with real numerical range merit special attention. Such elements, called Hermitian elements, possess many of the properties of a self-adjoint element of a B*-algebra, and there is a close connection between the two concepts. This chapter provides a systematic account of the connection.

In §5 we give two useful characterizations of Hermitian elements in a complex unital Banach algebra and then derive some elementary properties of the set of all Hermitian elements. The main result (known as Vidav's lemma) is that the numerical range of a Hermitian element is the convex hull of its spectrum. Vidav's lemma appears to be one of the deeper results in the present theory and we present various applications. The discussion of Hermitian elements is continued in §6 to lead up to the Vidav-Palmer theorem which gives a characterization of unital B*-algebras as those complex unital Banach algebras A such that each $a \in A$ is of the form $a = h + ik$ with h, k Hermitian elements.

The Vidav-Palmer theorem is used in §7 to give various characterizations of B*-algebras and also to give simple proofs of some known results in the theory of B*-algebras. In §8 we present mild generalizations of the Vidav-Palmer theorem to the cases of incomplete and non-unital algebras. A characterization of A*-algebras is immediately deduced. Finally we discuss the existence of a maximal B*-subalgebra of an arbitrary complex unital Banach algebra.

5. VIDAV'S LEMMA AND APPLICATIONS

Throughout this section, A denotes a complex unital Banach algebra. It will be clear that several of the results hold without the assumption of completeness, but the extensions to the incomplete case are usually trivial, since the numerical range is unchanged on passing to the completion. Moreover, we shall make frequent use of the exponential function, and some form of completeness is necessary for this.

Definition 1. We say that $h \in A$ is <u>Hermitian</u> if $V(A, h) \subset \underset{\sim}{R}$. We denote the set of all Hermitian elements of A by $H(A)$.

Lemma 2. Given $h \in A$, the following statements are equivalent.
  (i) $h \in H(A)$.
  (ii) $\lim_{\alpha \to 0} \frac{1}{\alpha} \{ \| 1 + i\alpha h \| - 1 \} = 0$.
  (iii) $\| \exp(i\alpha h) \| = 1 \quad (\alpha \in \underset{\sim}{R})$.

Remark. In (ii) above, and throughout this section, $\lim_{\alpha \to 0}$ denotes the limit as α approaches zero through <u>real</u> values.

Proof. We have $h \in H(A)$ if and only if

$$\max \operatorname{Re} V(A, ih) = \max \operatorname{Re} V(A, -ih) = 0.$$

Now apply Theorems 2.5 and 3.4.

Remark. Condition (ii) above was Vidav's original definition of Hermitian in [72]. A simple application of Lemma 2. 7 shows that $h \in A$ is Hermitian with respect to some equivalent algebra-norm on A if and only if the group $\{\exp(i\alpha h): \alpha \in \underline{R}\}$ is bounded. Lumer [42] calls such elements Hermitian-equivalent.

Example 3. Let A be a complex unital B*-algebra. Then $h \in H(A)$ if and only if $h^* = h$.

Proof. Let $h \in A$ with $h^* = h$, and let $\alpha \in \underline{R}$. We have

$$\|1 + \alpha^2 h^2\| = \|(1 + i\alpha h)(1 - i\alpha h)\| = \|1 + i\alpha h\|^2 ,$$

and therefore

$$\lim_{\alpha \to 0} \frac{1}{\alpha} \{\|1 + i\alpha h\| - 1\} = \lim_{\alpha \to 0} \frac{1}{\alpha} \{\|1 + \alpha^2 h^2\|^{\frac{1}{2}} - 1\} = 0.$$

This gives $h \in H(A)$.

On the other hand, given $a \in H(A)$, we have $a = u + iv$ with $u^* = u$, $v^* = v$, and

$$V(a) = \{f(u) + if(v): f \in D(A, 1)\} .$$

Since $a, u, v \in H(A)$, we have $f(v) = 0$ for all $f \in D(A, 1)$. Therefore $V(v) = \{0\}$, and so $v = 0$. The proof is complete.

Lemma 4. $H(A)$ is a real Banach space and $i(hk - kh) \in H(A)$ whenever $h, k \in H(A)$.

Proof. It is clear from the definition that $H(A)$ is a closed real linear subspace of A. Let $h, k \in H(A)$, and let $\alpha \in \underset{\sim}{R}$. It follows from Lemma 2 that $\| a \exp(ib) \| = \| a \|$ whenever $a \in A$, $b \in H(A)$. Therefore

$$\| \exp(i\alpha h)\exp(i\alpha k)\exp(-i\alpha h)\exp(-i\alpha k) \| = 1.$$

On expanding the exponentials we deduce that

$$\| 1 - \alpha^2 (hk - kh) \| = 1 + O(\alpha^3) \qquad (\alpha \to 0) \ .$$

With h, k interchanged we obtain

$$\| 1 + \alpha^2 (hk - kh) \| = 1 + O(\alpha^3) \qquad (\alpha \to 0) \ .$$

Therefore

$$\| 1 + \alpha(hk - kh) \| = 1 + o(\alpha) \qquad (\alpha \to 0) \ ,$$

and so $\lim\limits_{\alpha \to 0} \dfrac{1}{\alpha} \{ \| 1 + i\alpha(i(hk - kh)) \| - 1 \} = 0$. Hence $i(hk - kh) \in H(A)$ by Lemma 2.

The mapping $(h, k) \to hk + kh$ for $h, k \in H(A)$ will be discussed in the next section.

Definition 5. We say that $k \in A$ is <u>positive</u> if $V(A, k) \subset \underset{\sim}{R}^+$. We write $K(A)$ for the set of all positive elements of A.

Lemma 6. In the real Banach space $H(A)$, the set $K(A)$ is a normal closed cone in which 1 is an interior point.

Proof. Since $k \in K(A)$ if and only if

$$f(k) \geq 0 \quad (f \in D(A, 1)) ,$$

it is clear that $K(A)$ is a closed cone in $H(A)$. Given $h, k \in K(A)$, we have

$$\|h + k\| \geq v(h + k) \geq v(h) \geq e^{-1} \|h\| ,$$

so that $K(A)$ is a normal cone.

Finally let $h \in H(A)$ with $\|1 - h\| < 1$. Then $v(1 - h) < 1$, and so $V(h) \subset \underset{\sim}{R}^{+}$, $h \in K(A)$. This shows that 1 is an interior point of $K(A)$ in $H(A)$.

Notation. Let $J(A) = \{ h + ik : h, k \in H(A) \}$.

Lemma 7. Each element of $J(A)$ has unique representation of the form $h + ik$ with $h, k \in H(A)$.

Proof. Since $H(A)$ is a real linear space, it is enough to show that, if $h, k \in H(A)$ and $h + ik = 0$, then $h = k = 0$. The conditions on h, k give

$$V(h) = -iV(k) \subset \underset{\sim}{R} \cap i\underset{\sim}{R} = \{ 0 \} ,$$

so that, by Theorem 4.1, $h = 0$ and $k = 0$, as required.

By lemma 7, we may now define a mapping $*$ from $J(A)$ to itself by

$$(h + ik)* = h - ik \quad (h, k \in H(A)) .$$

It is trivial to verify that * is a linear involution on J(A), i. e.
$(a + b)^* = a^* + b^*$, $(\lambda a)^* = \lambda^* a^*$, $(a^*)^* = a$ whenever
$a, b \in J(A)$, $\lambda \in \underset{\sim}{C}$ (where λ^* denotes the complex conjugate
of λ).

Lemma 8. J(A) with the norm of A is a complex Banach
space, and * is a continuous linear involution on J(A).

Proof. Let $h, k \in H(A)$. Since h, k have real numerical
range, we have

$$v(h) \leq v(h + ik) \leq \|h + ik\| .$$

Theorem 4. 1 now gives

$$\|h\| \leq e\|h + ik\| , \tag{1}$$

and similarly $\|k\| \leq e\|h + ik\|$. It follows that

$$\|(h + ik)^*\| \leq \|h\| + \|k\| \leq 2e\|h + ik\| ,$$

so that the mapping * is continuous. Clearly J(A) is a complex
linear space. Let $\{h_n + ik_n\}$ be a Cauchy sequence in J(A). It
follows from (1) above that $\{h_n\}$, $\{k_n\}$ are Cauchy sequences in
H(A), and so converge by Lemma 4. Therefore $\{h_n + ik_n\}$
converges, as required.

The main result of this section (Theorem 10 below) has
come to be called Vidav's lemma. Vidav's original proof made use
of a fine version of the Phragmen-Lindelöf theorem (see Polya and
Szegö [55] Problem 325). We are indebted to G. Lumer for assis-

tance in simplifying[†] Vidav's proof. We make use of the following simple version of the Phragmén-Lindelöf theorem.

Lemma 9. Let f be an entire function such that
(i) for some $M \geq 0$, $|f(\zeta)| \leq M$ $(\mathrm{Re}\,\zeta \geq 0)$
(ii) $|f(\zeta)| \leq 1$ $(\mathrm{Re}\,\zeta = 0)$.
Then $|f(\zeta)| \leq 1$ $(\mathrm{Re}\,\zeta \geq 0)$.

Proof. Given $\epsilon > 0$, let

$$f_\epsilon(\zeta) = \frac{f(\zeta)}{1 + \epsilon\,\zeta} \quad (\mathrm{Re}\,\zeta \geq 0) .$$

Then $|f_\epsilon(\zeta)| \leq 1$ $(\mathrm{Re}\,\zeta = 0)$, and with $\mathrm{Re}\,\zeta \geq 0$ we have

$$\lim_{|\zeta| \to \infty} f_\epsilon(\zeta) = 0 .$$

Given ζ with $\mathrm{Re}\,\zeta \geq 0$, apply the maximum modulus principle to f_ϵ on a large semicircle with diameter on the imaginary axis to give $|f_\epsilon(\zeta)| \leq 1$. Let $\epsilon \to 0$ and we have

$$|f(\zeta)| \leq 1 \quad (\mathrm{Re}\,\zeta \geq 0) ,$$

as required.

Recall from Theorem 1.6 that a Hermitian element has real spectrum.

Theorem 10. (Vidav [72]). Let $h \in H(A)$. Then

$$\max V(A, h) = \max Sp(A, h) .$$

---

† F. F. Bonsall and M. J. Crabb (preprint) have now given an even simpler proof.

Proof. Let $\mu = \max V(A, h)$, $\sigma = \max Sp(A, h)$, and let ϕ be defined on $\underset{\sim}{R}^+$ by

$$\phi(\xi) = \log \|\exp(\xi h)\| \qquad (\xi \geq 0) \ .$$

Theorems 3. 4 and 3. 8 give

$$\mu = \sup \{ \xi^{-1} \phi(\xi) : \xi > 0 \} \ , \tag{2}$$

$$\sigma = \inf \{ \xi^{-1} \phi(\xi) : \xi > 0 \} \tag{3}$$

$$= \lim_{\xi \to +\infty} \xi^{-1} \phi(\xi) \ .$$

Given $\epsilon > 0$, there exists $\xi_0 > 0$ such that

$$\xi^{-1} \phi(\xi) \leq \sigma + \epsilon \qquad (\xi > \xi_0) \ ,$$

and so

$$\|\exp(\xi h)\| \leq \exp((\sigma + \epsilon)\xi) \qquad (\xi > \xi_0) \ .$$

Hence there exists $K_\epsilon > 0$ such that

$$\|\exp(\xi h)\| \leq K_\epsilon \exp((\sigma + \epsilon)\xi) \qquad (\xi \geq 0) \ . \tag{4}$$

Given $g \in A'$ with $\|g\| \leq 1$, let f be defined on $\underset{\sim}{C}$ by

$$f(\zeta) = g(\exp(-(\sigma + \epsilon)\zeta)\exp(\zeta h)) \qquad (\zeta = \xi + i\eta) \ .$$

Then f is an entire function and

$$|f(\zeta)| \leq \|\exp(-(\sigma + \epsilon)\zeta)\exp(\xi h)\exp(i\eta h)\|$$

$$\leq \exp(-(\sigma + \epsilon)\xi)\|\exp(\xi h)\| .$$

Thus $|f(\zeta)| \leq 1$ $(\text{Re}\zeta = 0)$, and by (4)

$$|f(\zeta)| \leq K_\epsilon \qquad (\text{Re}\zeta \geq 0) .$$

It follows from Lemma 9 that $|f(\zeta)| \leq 1$ $(\text{Re}\zeta \geq 0)$ and, since g was arbitrary, this gives

$$\|\exp(-(\sigma + \epsilon)\zeta)\exp(\zeta h)\| \leq 1 \qquad (\text{Re}\zeta \geq 0) ,$$

and therefore

$$\|\exp(\xi h)\| \leq \exp((\sigma + \epsilon)\xi) \qquad (\xi \geq 0) .$$

Let $\epsilon \to 0$, and we obtain

$$\|\exp(\xi h)\| \leq \exp(\sigma \xi) \qquad (\xi \geq 0) .$$

It follows from (3) that

$$\|\exp(\xi h)\| = \exp(\sigma \xi) \qquad (\xi \geq 0) , \qquad\qquad (5)$$

and so $\phi(\xi) = \sigma \xi$ $(\xi \geq 0)$. Then $\mu = \sigma$ by (2).

Corollary 11. $V(h) = \text{co Sp}(h)$ $(h \in H(A))$.

Proof. We have

$$\text{minV}(h) = -\text{maxV}(-h) = -\text{maxSp}(-h) = \text{minSp}(h)$$

and the result follows since $V(h)$ is compact convex.

Corollary 12. $\|h\| \le e\rho(h)$ $(h \in H(A))$.

Proof. Apply Corollary 11 and Theorem 4.1.

Remark. Sinclair [62] has recently proved the remarkable theorem that

$$\rho(h) = \|h\| \qquad (h \in H(A)) . \tag{6}$$

The proof again turns on a Phragmén-Lindelöf argument, and of course Vidav's lemma is an immediate corollary of (6). The weaker result

$$\rho(\exp(h)) = \|\exp(h)\| \qquad (h \in H(A)) \tag{7}$$

follows immediately from the proof of Vidav's lemma. In fact (5) gives

$$\|\exp(nh)\| = \exp(n\sigma) = \|\exp(h)\|^n \qquad (n = 1, 2, \dots) \tag{8}$$

and (7) follows.

Definition 13. We say that $a \in A$ is <u>normal</u> if $a = h + ik$ with $h, k \in H(A)$, $hk = kh$.

Theorem 14. Let $a \in A$ be normal. Then

$$V(A, a) = \text{co } Sp(A, a) .$$

Proof. We have co $Sp(A, a) \subset V(A, a)$ by Theorems 2.3 and 2.6. Let $a = h + ik$, where $h, k \in H(A)$, $hk = kh$, let B be a maximal commutative subalgebra of A containing h and k and let Φ be the maximal ideal space of B. Then

$$Sp(A, a) = Sp(B, a) = \hat{a}(\Phi) = \{\hat{h}(\phi) + i\hat{k}(\phi): \phi \in \Phi\}$$

and $Sp(A, h) = Sp(B, h) = \hat{h}(\Phi)$, $Sp(A, k) = Sp(B, k) = \hat{k}(\Phi)$. Since $h, k \in H(A)$, we have $Sp(A, h) \subset \underset{\sim}{R}$, $Sp(A, k) \subset \underset{\sim}{R}$, by Theorem 2.6. Suppose that Re $Sp(A, a) \subset \underset{\sim}{R}^+$. Then $\hat{h}(\Phi) \subset \underset{\sim}{R}^+$ and so $Sp(A, h) \subset \underset{\sim}{R}^+$. Therefore $V(A, h) \subset \underset{\sim}{R}^+$ by Theorem 10, and so Re $V(A, a) \subset \underset{\sim}{R}^+$. Since $\lambda 1 + \mu a$ is normal for $\lambda, \mu \in \underset{\sim}{C}$, and since co $Sp(A, a)$ is the intersection of all the closed half-planes that contain $Sp(A, a)$, the result now follows.

We are indebted to G. Lumer for the following improvement of the result mentioned after Theorem 4.1.

Theorem 15. Let A be a real unital Banach algebra, and let $a \in A$ with $V(A, a) = \{0\}$. Then $\|a\| \leq e\rho(a)$. If, further, $V(A, a^2) = \{0\}$, then $a = 0$.

Proof. Let B be any complexification of A, so that A is embedded isometrically in B. By Theorem 2.5 we have

$\lim_{\alpha \to 0} \alpha^{-1}\{\|1 + \alpha a\| - 1\} = 0$ in A, and so in B. Therefore, by Lemma 2, $a = ih$ for some $h \in H(B)$. Since $Sp(A, a) = Sp(B, a)$ (by definition), it follows from Corollary 12 that $\|a\| \leq e\rho(a)$. Now let $V(A, a^2) = \{0\}$. Then $a^2 = ik$ for some $k \in H(B)$. Since $a^2 = -h^2$, this gives $Sp(B, h^2) \subset i\underset{\sim}{R}$ and so $Sp(B, h^2) = \{0\}$. Therefore $Sp(B, h) = \{0\}$ and Corollary 12 gives $h = 0$, $a = 0$.

The result below due to Spatz [64] may be proved by elementary arguments. It is also an amusing application of the uses of Hermitian elements.

Theorem 16. (Spatz.) Let A be a real or complex unital Banach algebra such that $\lim\limits_{\alpha \to 0} \alpha^{-1} \{ \|1 + \alpha x\| - 1 \}$ exists for each singular element x. Then A is a division algebra.

Proof. Let $x \in A$ be singular, and let $\lim\limits_{\alpha \to 0} \alpha^{-1} \{ \|1 + \alpha x\| - 1 \} = \lambda$. If A is complex, then $x = \lambda + ih$ for some $h \in H(A)$. Since $0 \in Sp(x)$, $Sp(h) \subset \underline{R}$, we have $\lambda = 0$, $x = ih$. Since ix is also singular, we have $ix = ik$ for some $k \in H(A)$ and therefore $x = H(A) \cap iH(A)$, $x = 0$. If A is real, then, by Theorem 2. 6,

$$0 \in Sp(x) \cap \underline{R} \subset V(x) = \{\lambda\} ,$$

and so $V(x) = \{0\}$. Since x^2 is also singular, we have $V(x^2) = \{0\}$, and therefore $x = 0$ by Theorem 15.

6. THE VIDAV-PALMER THEOREM

A C*-algebra is a uniformly closed self-adjoint subalgebra of $B(\mathscr{H})$ for some complex Hilbert space $\mathscr{H}$. Gelfand and Naimark [24] gave an abstract characterization of such algebras amongst the class of Banach star algebras, and we shall discuss their results in the next section. The first step to a characterization of C*-algebras in the class of all complex Banach algebras was taken by Vidav [72] who proved the following result.

"Let A be a complex unital Banach algebra such that

(i) $A = H(A) + iH(A)$,

(ii)   if $h \in H(A)$, then h^2 is normal.

Then there is a bicontinuous isomorphism $a \to T_a$ of A with a
C*-algebra such that T_h is self-adjoint and $|T_h| = \|h\|$ whenever
$h \in H(A)$. "

Berkson [8] and Glickfeld [27] showed independently (and
by different methods) that the mapping $a \to T_a$ is in fact an
isometry. Finally, Palmer [49] showed that condition (ii) is
unnecessary and he also gave the simplest proof that $a \to T_a$
is an isometry. Example 5. 3 shows that unital C*-algebras
satisfy condition (i). Thus condition (i) is a complete abstract
characterization of unital C*-algebras. The proof of this result,
which we shall call the Vidav-Palmer theorem, is the main aim of
this section.

We give first some preliminaries about the Banach space
$J(A) = H(A) + iH(A)$ for an arbitrary complex unital Banach
algebra A. The Banach space $J(A)$ need not be a subalgebra of
A. Lumer [42] gave an example of a Hermitian equivalent operator
some power of which is not Hermitian equivalent. This implies the
existence of a Hermitian operator the square of which is not
Hermitian. The following example is due to M. J. Crabb. Let
$A = \underset{\sim}{C}^3$ with pointwise multiplication, and let p be defined on
$\underset{\sim}{C}^3$ by

$$p(\alpha, \beta, \gamma) = \sup\{\,|\lambda^{-1}\alpha + \beta + \lambda\gamma|: \lambda \in \underset{\sim}{C}, \ |\lambda| = 1\,\}\,.$$

Then p is a Banach space norm on A. Let $\|.\|$ be the corres-
ponding operator-norm on A, i. e.

$$\|a\| = \sup\{p(xa): x \in A, \; p(x) = 1\} \; ,$$

so that A is a complex unital (commutative) Banach algebra with
respect to $\|.\|$.

Example 1. Let $A = \underset{\sim}{C}^3$ with norm defined as above, and
let $h = (-1, 0, 1)$. Then $h \in H(A)$, $h^2 \notin H(A)$.

Proof. Given $t \in \underset{\sim}{R}$, $x \in A$, $x = (\alpha, \beta, \gamma)$, we have

$$p(x \exp(ith)) = p(\alpha \exp(-it), \; \beta, \; \gamma \exp(it)) = p(\alpha, \beta, \gamma) \; ,$$

so that $\|\exp(ith)\| = 1$. Therefore $h \in H(A)$, by Lemma 5. 2.
Let $s = -\frac{1}{2}\pi$, and $y = (i, 1, i)$. Then

$$p(y) \; \|\exp(ish^2)\| \geq p(y \exp(ish^2)) = p(1) \; .$$

Since $p(y) = 5^{\frac{1}{2}}$, $p(1) = 3$, it follows that $\|\exp ish^2\| > 1$, and
therefore, by Lemma 5. 2, $h^2 \notin H(A)$.

Corollary 2. Let A be as above, and let k, $k^2 \in H(A)$.
Then $k = \mu 1$ for some $\mu \in \underset{\sim}{R}$.

Proof. We note that A is the linear span of $1, h, h^2$. Let
$u \in H(A)$ where $u = \alpha + \beta h + \gamma h^2$. Then $Sp(A, u) \subset \underset{\sim}{R}$ and so

$$\{\alpha, \; \alpha + \beta + \gamma, \; \alpha - \beta + \gamma\} \subset \underset{\sim}{R} \; .$$

It follows that $\alpha, \beta, \gamma \in \underset{\sim}{R}$. Moreover $\gamma = 0$, for otherwise $h^2 \in H(A)$. We now have $H(A) = \{\alpha + \beta h : \alpha, \beta \in \underset{\sim}{R}\}$ and the corollary is clear.

Remark. For elements $h \in H(A)$ such that $h^n \in H(A)$ $(n = 2, 3, \dots)$, it turns out, via the Vidav-Palmer theorem, that

$$\|1 + ith\| = 1 + O(t^2) \text{ as } t \to 0 . \tag{1}$$

It is tempting to conjecture that the pathology of the above example might lie in the order of magnitude of $\|1 + ith\|$ as $t \to 0$. However, an elementary calculation shows that the element h of Example 1 also satisfies condition (1) above.

Recall from Lemma 5. 8 that the mapping * from $J(A)$ to itself is given by $(h + ik)* = h - ik$, and is a linear involution on $J(A)$.

Theorem 3. Let A be a complex unital Banach algebra. Then the following statements are equivalent.

(i) $J(A)$ is an algebra.
(ii)   If $h \in H(A)$, then $h^2 \in J(A)$.
(iii)   If $h \in H(A)$, then $h^2 \in H(A)$.
(iv)   If $h, k \in H(A)$, then $hk + kh \in H(A)$.

Under any of the above conditions, $J(A)$ is a Banach star algebra with continuous involution and with $H(A)$ as its set of self-adjoint elements.

Proof.   (i) $\to$ (ii) is trivial.
(ii) $\to$ (iii). Let $h \in H(A)$ with $h^2 = p + iq$, where

$p, q \in H(A)$. Then $h(p + iq) = (p + iq)h$, and so $hp - ph = i(qh - hq) = c$, say. By Lemma 5.4, we have $c, ic \in H(A)$ and therefore $V(c) = \{0\}$, $c = 0$. This gives $hp = ph$, $h^2 p = ph^2$, and so $pq = qp$. Therefore h^2 is normal and $V(h^2) = \text{co } Sp(h^2) = \text{co } Sp(h)^2 \subset \mathbb{R}$ by Theorem 5.14.

(iii) → (iv). Let condition (iii) hold. Given $h, k \in H(A)$, we have $h^2, k^2, (h + k)^2 \in H(A)$, and hence $hk + kh \in H(A)$.

(iv) → (i). Let condition (iv) hold. It will be enough to show that $J(A)$ is closed under multiplication. Let $a, b \in J(A)$, where $a = h + ik$, $b = p + iq$. Using Lemma 5.4 we now have

$$ab + b^*a^* = (hp + ph) - (kq + qk) + i(kp - pk) + i(hq - qh)$$
$$\in H(A)$$

$$i(ab - b^*a^*) = i(hp - ph) - i(kq - qk) - (kp + pk) - (hq + qh)$$
$$\in H(A) .$$

Therefore

$$ab = \tfrac{1}{2}(ab + b^*a^*) - i\tfrac{1}{2}i(ab - b^*a^*) \in J(A).$$

If $J(A)$ is an algebra, then

$$(ab)^* = \tfrac{1}{2}(ab + b^*a^*) + i\tfrac{1}{2}i(ab - b^*a^*) = b^*a^* ,$$

and Lemma 5.8 completes the proof.

The equivalence of (ii) and (iii) was proved by Palmer [49]. Example 1 shows that $H(A)$ may fail to be closed under the real Jordan product $hk + kh$, although $H(A)$ is always closed under the imaginary Jordan product $i(hk - kh)$.

Recall that $K(A) = \{k \in A : V(k) \subset \underset{\sim}{R}^+\}$. The purpose of the next four lemmas is to show that, if $J(A)$ is an algebra, then $J(A)$ is symmetric, i. e. $Sp(J(A), x^*x) \subset \underset{\sim}{R}^+$ whenever $x \in J(A)$. It is enough to prove that $x^*x \in K(A)$ whenever $x \in J(A)$. Since each self-adjoint element of $J(A)$ is Hermitian and so has real spectrum, it is an immediate consequence of the Shirali-Ford theorem [61] that $J(A)$ is symmetric. The method of proof used below is essentially due to Kaplansky. Lemma 4 is of independent interest and will be required later. It is a special case of the theorem of Sinclair [62] to which we have referred already.

Lemma 4. (Vidav.) Let $J(A)$ be an algebra. Then $\rho(h) = \|h\|$ for each $h \in H(A)$.

Proof. Let $h \in H(A)$. By Theorem 3 and a simple induction argument, we have $h^n \in H(A)$ for each positive integer n. Let C be the closed subalgebra of $J(A)$ generated by 1 and h. Since the involution on $J(A)$ is continuous, C is a star subalgebra of $J(A)$ and therefore each element of C is normal. Since

$$Sp(C, h) \subset V(C, h) = V(A, h) \subset \underset{\sim}{R},$$

we have $1 + i\xi h \in G(C)$ for $\xi \in \underset{\sim}{R}$. Since the mapping $t \to 1 + ith$ $(t \in \underset{\sim}{R})$ is continuous, it follows that, for $\xi \in \underset{\sim}{R}$, $1 + i\xi h \in G_1(C)$, the principal component of $G(C)$. By Rickart [57] Corollary (1.4.11), there is $w \in C$ such that $1 + i\xi h = \exp(w)$. We have $w = u + iv$, where $u, v \in H(C)$, $uv = vu$. It is clear that $\exp(u)^* = \exp(u)$, $\exp(iv)^* = \exp(-iv)$.

c

Therefore

$$1 + i\xi h = \exp(u) \exp(iv), \quad 1 - i\xi h = \exp(u) \exp(-iv) .$$

Using formula (8) of §5, we now have

$$\| 1 + \xi^2 h^2 \| = \| \exp(2u) \| = \| \exp(u) \|^2 = \| 1 + i\xi h \|^2 \quad (\xi \in \mathbb{R}) .$$

Let $\xi \to +\infty$, and we deduce that $\| h^2 \| = \| h \|^2$. Since $h^n \in H(A)$ for $n = 1, 2, \ldots$, it follows that $\rho(h) = \| h \|$.

Remark. The above technique may be used to show that if $J(A)$ is an algebra then $\| x^*x \| = \| x \|^2$ whenever x is normal and invertible.

Lemma 5. Let $J(A)$ be an algebra. Given $k \in K(A)$, let C denote the least closed subalgebra of A containing 1 and k. Then there exists $u \in C \cap K(A)$ such that $u^2 = k$.

Proof. We may clearly suppose that $\rho(k) \le 1$. Let $a = 1 - k$ and let P denote the set of polynomials in $1, a$ with coefficients in $\mathbb{R}^+$. Let

$$x_0 = 0, \quad x_n = \tfrac{1}{2}(a + x_{n-1}^2) \quad (n = 1, 2, \ldots) .$$

By induction, $x_n \in P$, and since

$$x_{n+1} - x_n = \tfrac{1}{2}(x_n + x_{n-1})(x_n - x_{n-1}) ,$$

we have $x_n - x_{n-1} \in P$ $(n = 1, 2, \ldots)$. Further, since $ax_n = x_n a$, it follows that $\rho(x_n) \le 1$ $(n = 1, 2, \ldots)$.

Let $x \to \hat{x}$ be the Gelfand representation of C on its maximal ideal space Φ_C. Since $Sp(C, a) \subset [0, 1]$, $\{\hat{x}_n\}$ is an increasing sequence of non-negative functions on Φ_C bounded above by 1 and so pointwise convergent to a function w with $0 \leq w \leq 1$ on Φ_C. Since $w = \frac{1}{2}(\hat{a} + w^2)$ and $\hat{a}$ is continuous, w is continuous, and so, by Dini's theorem, $\hat{x}_n \to w$ uniformly on Φ_C. Therefore $\{x_n\}$ is a Cauchy sequence with respect to ρ, and so, by Lemma 4, with respect to $\|.\|$. Therefore $x_n \to v \in H(A)$, $\hat{v} = w$. Let $u = 1 - v$, and we have $u^2 = k$, and $Sp(A, u) \subset Sp(C, u) \subset [0, 1]$. Vidav's lemma gives $u \in K(A)$, and the proof is complete.

Lemma 6. Let $J(A)$ be an algebra. Then, for each $h \in H(A)$, there exist $p, q \in K(A)$ such that $h = p - q$, $pq = qp = 0$.

Proof. By Lemma 5, there is $|h| \in K(A)$ such that $|h|^2 = h^2$. Let C be the closed subalgebra of $J(A)$ generated by 1 and h, so that $|h| \in C$. Let Φ be the maximal ideal space of C, and then

$$Sp(C, |h| \pm h) = \{\phi(|h|) \pm \phi(h) : \phi \in \Phi\} .$$

Since $h^2 = |h|^2$, we have, for each $\phi \in \Phi$,

$$\phi(h)^2 = \phi(h^2) = \phi(|h|^2) = \phi(|h|)^2 ,$$

and therefore $\phi(h) = \pm \phi(|h|)$. Since h, $|h| \in H(C)$, it follows that $Sp(C, |h| \pm h) \subset \mathbb{R}^+$ and so $|h| \pm h \in K(C)$ by Vidav's lemma. Then $|h| \pm h \in K(A)$, and if $p = \frac{1}{2}(|h| + h)$, $q = \frac{1}{2}(|h| - h)$, we have $h = p - q$, $p, q \in K(A)$, $pq = qp = 0$.

Let $J(A)$ be an algebra and let $x \in J(A)$ with $x = h + ik$. Then

$$x^*x = h^2 + k^2 + i(hk - kh) \, ,$$

so that $x^*x \in H(A)$. The next lemma shows that x^*x is in fact positive.

Lemma 7. Let $J(A)$ be an algebra and let $x \in J(A)$. Then $x^*x \in K(A)$.

Proof. Let $y \in J(A)$ be such that $-y^*y \in K(A)$. Since

$$Sp(A, y^*y) \setminus \{0\} = Sp(A, yy^*) \setminus \{0\} \, ,$$

it follows from Vidav's lemma that $-yy^* \in K(A)$. Let $y = h + ik$ where $h, k \in H(A)$. Then $y^*y = 2h^2 + 2k^2 - yy^* \in K(A)$, since $K(A)$ is a cone. Therefore $V(y^*y) = \{0\}$ and so $y^*y = 0$. Similarly $yy^* = 0$, and therefore $h^2 + k^2 = 0$, $h = k = 0$, $y = 0$.

Suppose now that x is any element of $J(A)$. Then $x^*x \in H(A)$, and so, by Lemma 6, there exist $p, q \in K(A)$ such that $x^*x = p - q$, $pq = qp = 0$. Let $y = xq$. Then $-y^*y = -qx^*xq = q^3 \in K(A)$, and therefore $y = 0$ by the above paragraph. This gives in turn, $x^*xq = 0$, $q^2 = 0$, $q = 0$, $x^*x = p \in K(A)$.

We need one more technical lemma before the main theorem. The lemma is an extension by Palmer [49] of a result of Russo and Dye [58], and we state it without proof. An elementary proof can be based on L. A. Harris, 'Schwarz's lemma in normed linear spaces', Proc. Nat. Acad. Sci. 62 (1969) 1014-1017.

Lemma 8. Let A be a unital C*-algebra and let $E = \{\exp(ih): h \in A, h^* = h\}$. Then the closed unit ball of A is the closed convex hull of E.

Theorem 9. (Vidav-Palmer.) Let A be a complex unital Banach algebra such that $A = H(A) + iH(A)$. Then A is isometrically star isomorphic with a C*-algebra.

Proof. Since $J(A) = A$, certainly A is a Banach star algebra and Lemma 7 applies. Let $f \in D(A, 1)$ and let $x \in A$, where $x = h + ik$, $h, k \in H(A)$. Then $f(x^*x) \geq 0$ and

$$f(x^*) = f(h) - if(k) = f(x)^* .$$

Therefore each $f \in D(A, 1)$ is a positive Hermitian functional. By Rickart [57] Theorem (4.5.4) and Lemma 5.8, for each $f \in D(A, 1)$ there is a Hilbert space $\mathcal{H}_f$ and a star homomorphism $a \to T_a^f$ from A into $B(\mathcal{H}_f)$ such that

$$|T_h^f| \leq \|h\| \qquad (h \in H(A))$$

$$|T_a^f| \leq (2e)^{\frac{1}{2}} \|a\| \qquad (a \in A) .$$

Let $a \to T_a$ denote the star homomorphism from A into $B(\mathcal{H})$ obtained by forming the direct sum of all star homomorphisms associated with elements of $D(A, 1)$. Then

$$|T_h| \leq \|h\| \qquad (h \in H(A)) ,$$

$$|T_a| \leq (2e)^{\frac{1}{2}} \|a\| \qquad (a \in A) ,$$

and

$$|T_a|^2 \geq \sup\{f(a*a) : f \in D(A, 1)\}$$

$$= v(a*a) = \|a*a\|^{\frac{1}{2}} \qquad (a \in A) ,$$

by Lemmas 4 and 7. By applying the Cauchy-Schwarz inequality for each $f \in D(A, 1)$, we obtain

$$v(a)^2 \leq v(a*a) ,$$

and so, by Theorem 4.1,

$$e|T_a| \geq \|a\| \qquad (a \in A) .$$

It follows that $a \to T_a$ is one-to-one, has a C*-algebra as range and satisfies $|T_h| = \|h\|$ $(h \in H(A))$.

By identifying A with its range under $a \to T_a$, we now suppose that A is a C*-algebra with C*-norm $|.|$ and second norm $\|.\|$. Let $E = \{\exp(ih): h \in H(A)\}$. Given $a \in A$ with $|a| = 1$, by Lemma 8, there is a sequence $\{b_n\}$ in $co(E)$ such that $|b_n - a| \to 0$. Therefore $\|b_n - a\| \to 0$ and so $\|b_n\| \to \|a\|$. By Lemma 5.2, $\|\exp(ih)\| = 1$ $(h \in H(A))$ and therefore $\|b_n\| \leq 1$, $n = 1, 2, \ldots$, from which $\|a\| \leq 1$. It follows that $\|a\| \leq |a|$ $(a \in A)$. If for some $b \in A$ we have $\|b\| < |b|$, then

$$\|b*b\| \leq \|b*\|\,\|b\| < |b*|\,|b| = |b*b| = \|b*b\| .$$

This contradiction completes the proof.

The strongly geometrical nature of the Vidav-Palmer theorem may be emphasized as follows. If A is a complex unital Banach algebra, then the existence of an involution on A for which A is a B*-algebra depends only on the Banach space structure of A, in fact on the shape of the unit ball near 1. From a slightly different viewpoint, suppose that B is a complex Banach space. Let $u \in S(B)$, $H(B) = \{a \in B: f(a) \in \mathbb{R} \ (f \in D(B, u) \}$, and suppose that $B = H(B) + iH(B)$. If o is any multiplication on B such that (B, o) is a Banach algebra with unit element u, then (B, o) is a B*-algebra. It would be of interest to characterize those complex Banach spaces that admit a B*-algebra structure.

7. APPLICATIONS OF THE VIDAV-PALMER THEOREM TO B*-ALGEBRAS

Throughout this section A will denote a complex Banach star algebra. We say that A is a B*-algebra if it satisfies the condition

$$\|x^*x\| = \|x\|^2 \qquad (x \in A) . \tag{1}$$

We say that A is a B_0^*-algebra if it satisfies the (apparently) weaker condition

$$\|x^*x\| = \|x^*\| \ \|x\| \qquad (x \in A) . \tag{2}$$

We show in this section how the Vidav-Palmer theorem may be used to give various characterizations of B*-algebras amongst Banach star algebras, and show that it leads to simple proofs of

various known results in the theory of B*-algebras. We begin
however with a discussion of the classical Gelfand-Naimark
result that B*-algebras are merely abstract C*-algebras.

Theorem 1. Let A be a B*-algebra. Then A is
isometrically star isomorphic with a C*-algebra.

Proof. We may suppose by Rickart [57] Lemma (4. 1. 13)
that A is unital. Example 5. 3 shows that the Hermitian elements
of A coincide with the self-adjoint elements. The Vidav-Palmer
theorem now applies.

Remark. It should be noted that the full force of the
Vidav-Palmer theorem is not required for the above theorem. It
is trivial to show that $\rho(h) = \|h\|$ for h self-adjoint. We then
need Lemmas 6. 5, 6. 6, 6. 7 as before. However, in the proof of
the Vidav-Palmer theorem, condition (1) gives immediately that
$|T_a| = \|a\|$ (a $\in$ A). This would appear to be the most economical
proof of the Gelfand-Naimark theorem.

The sharper form of the Gelfand-Naimark theorem below
was proved in the unital case by Glimm and Kadison [29]. The case
when A has no unit element was announced by Ono [48]. Vowden
[73] has recently given a complete proof of the result based on
Ono's method.

Theorem 2. Let A be a B_o^*-algebra. Then A is
isometrically star isomorphic with a C*-algebra.

Proof. We may suppose by Vowden [73] that A is unital.
As in the proof of Theorem 1, it will be sufficient to show that

each self-adjoint element of A is Hermitian. Let $h \in A$ with $h^* = h$, let $\mu = \max \mathrm{Re}\ V(ih)$ and $\lambda = \min \mathrm{Re}\ V(ih)$. For $\alpha > 0$, Theorem 2.5 gives

$$\|1 + i\alpha h\| = 1 + \alpha\mu + o(\alpha), \quad \|1 - i\alpha h\| = 1 - \alpha\lambda + o(\alpha) .$$

Therefore

$$
\begin{aligned}
1 + \alpha(\mu - \lambda) + o(\alpha) &= \|1 + i\alpha h\|\ \|(1 + i\alpha h)^*\| \\
&= \|1 + \alpha^2 h^2\| = 1 + O(\alpha^2) .
\end{aligned}
$$

It follows that $\lambda = \mu$, $V(ih) \subset \lambda + i\mathbb{R}$, $V(h + i\lambda 1) \subset \mathbb{R}$, and so there is $k \in H(A)$ such that $h = k - i\lambda 1$. Since $h^* = h$, $\mathrm{Sp}(h)$ is self-conjugate, and since $k \in H(A)$, $\mathrm{Sp}(k) \subset \mathbb{R}$. It follows that $\lambda = 0$ and therefore $h \in H(A)$, as required.

Remark. It is clear from the above proof that the full force of condition (2) is not required. If A is a unital Banach star algebra such that

$$\frac{\|x^*x\|}{\|x\|\ \|x^*\|} = 1 + o(r), \quad r = \|1 - x\|, \quad r \to 0 ,$$

then the above proof shows that A is a B*-algebra. This "local differential" condition for a B*-algebra is due to Lumer [40].

A linear functional f on A is said to be <u>Hermitian</u> if

$$f(x^*) = f(x)^* \quad (x \in A) .$$

Clearly f is Hermitian if and only if f takes real values on self-adjoint elements.

Theorem 3. Let A be a unital Banach star algebra such that $f \in A'$ is Hermitian whenever $f(1) = \|f\|$. Then A is a B*-algebra.

Proof. Let $h \in A$ with $h^* = h$. Then

$$V(h) = \{f(h) : f(1) = \|f\| = 1\} \subset \mathbb{R}$$

and Vidav-Palmer applies.

Our final characterization of B*-algebras is essentially "spectral".

Theorem 4. Let A be a unital Banach star algebra such that

(i) $Sp(A, h) \subset \mathbb{R}$ when $h^* = h$,

(ii) $\rho(1 + \lambda h) = \|1 + \lambda h\|$ when $h^* = h$, $\lambda \in \mathbb{C}$.

Then A is a B*-algebra.

Proof. Let $h^* = h$. Condition (ii), together with the characterization of co Sp(h) as the intersection of the closed discs that contain Sp(h), shows that $V(h) = \text{co } Sp(h)$. Condition (i) now gives $V(h) \subset \mathbb{R}$ and then Vidav-Palmer applies.

It is tempting to weaken condition (ii) in Theorem 4 to hold for real scalars only. But then A may fail to be a B*-algebra, even in the commutative case. This is shown by the example below.

Example 5. Let E be a compact Hausdorff space with at least two points and let $A = C(E)$, the set of all continuous complex functions on E. Let $\|.\|_o$ be defined on A by

$$\|f\|_o = \sup \{\tfrac{1}{2}|f(s) + f(t)| + \tfrac{1}{2}|f(s) - f(t)| : s, t \in E\} .$$

Then $(A, \|.\|_o)$ is a unital Banach algebra and $\|.\|_o$ coincides with the supremum norm on real functions but not on all of A.

Proof. Let $\|.\|_\infty$ denote the supremum norm on A. It is straightforward to verify that $\|.\|_o$ is an algebra-norm on A such that

$$\|f\|_\infty \le \|f\|_o \le 2\|f\|_\infty \qquad (f \in A).$$

For real a, b we have

$$\max(|a|, |b|) = \tfrac{1}{2}|a + b| + \tfrac{1}{2}|a - b|$$

and therefore $\|f\|_o = \|f\|_\infty$ for real valued functions. Finally, it is easy to construct $g \in A$ for which $\|g\|_\infty < \|g\|_o$.

On the other hand, there is a unique absolute algebra-norm on $C(E)$ that is normalized at 1.

Theorem 6. Let E be a compact Hausdorff space and let $A = C(E)$. Let $\|.\|$ be a Banach algebra-norm on A such that $\|1\| = 1$ and $\|f\| = \| |f| \|$ $(f \in A)$. Then $\|.\|$ is the supremum norm.

Proof. Let g be a continuous real function on E and let $t \in \underset{\sim}{R}$. Then

$$\|\exp(itg)\| = \| |\exp itg| \| = \|1\| = 1 ,$$

71

and so $g \in H(A)$, by Lemma 5.2. The Vidav-Palmer theorem now shows that A is a B*-algebra with the natural involution and hence $\|.\|$ is the spectral norm, i.e. the supremum norm.

Theorem 7. (Segal [59] and Kaplansky [37].) Let A be a B*-algebra and let I be a closed two-sided ideal of A. Then $I^* = I$, and A/I is a B*-algebra.

Proof. We write $[a]$ for the canonical image of a in A/I. Suppose first that A is unital, so that A/I is unital, with unit element $[1]$. Given $\phi \in D(A/I, [1])$, let $\tilde{\phi}$ be defined on A by

$$\tilde{\phi}(a) = \phi([a]) \qquad (a \in A) .$$

Then $\tilde{\phi} \in D(A, 1)$. Let $h \in A$ with $h^* = h$, so that $h \in H(A)$. Then $\phi([h]) = \tilde{\phi}(h) \in R$ and so $[h] \in H(A/I)$. Since $A/I = \{[h] + i[k] : h, k \in H(A) \}$, it follows from the Vidav-Palmer theorem that A/I is a B*-algebra with the canonical norm and involution $([h] + i[k])^* = [h] - i[k]$. Therefore $a \to [a]$ is a star homomorphism and so $I^* = I$.

If A does not have a unit element, we may adjoin one to form a unital B*-algebra B, say (Rickart [57] Lemma (4.1.13)). Then I is a closed two-sided ideal of B, so that $I^* = I$ and B/I is a B*-algebra, by the above paragraph. Since A/I is a closed star subalgebra of B/I, the proof is complete.

The theorem below characterizes linear isometries between B*-algebras and is due to Kadison [35]. Lumer [40] gave a numerical range proof of the commutative case, and more recently,

Paterson [53] has given a numerical range proof of the full theorem.

> **Theorem 8.** (Kadison [35], Paterson [53].) Let A, B be unital B*-algebras and let T be a linear isometry from A onto B. Then there is a C*-isomorphism S from A onto B, and a unitary element u of B such that T = uS.

A further application of the Vidav-Palmer theorem is given in §12, where it is used to show that the second dual of a B*- algebra is again a B*-algebra with the Arens multiplication and the natural involution.

8. OTHER APPLICATIONS OF THE VIDAV-PALMER THEOREM

Let A be a complex Banach star algebra. Then A is said to be an A*-algebra if it has a second (or auxiliary) algebra-norm $|.|$ such that

$$|x^*x| = |x|^2 \qquad (x \in A) .$$

The norm $|.|$ need not be a complete norm on A. It is convenient to have a mild generalization of A*-algebras.

> **Definition 1.** A normed star algebra $(A, |.|)$ is said to be a pre-B*-algebra if

$$|x^*x| = |x|^2 \qquad (x \in A) .$$

It is clear that the completion of a pre-B*-algebra A is a
B*-algebra, and therefore, by Theorem 2. 4 and Example 5. 3,
each self-adjoint element of A is Hermitian if A is also unital.
The theorem below is therefore a characterization of unital pre-
B*-algebras.

Theorem 2.　Let A be a complex unital normed algebra
such that $A = H(A) + iH(A)$.　Then A is a pre-B*-algebra.

Proof.　Let $|.|$ denote the norm on A, and let B be
the completion of A. We may identify $D(A, 1)$ with $D(B, 1)$.
Let M be the closure of $H(A)$ in B. Given $h \in M$, there exists
a sequence $\{h_n\}$ in $H(A)$ with $h = \lim_{n \to \infty} h_n$. For each
$f \in D(B, 1)$ we thus have

$$f(h) = \lim_{n \to \infty} f(h_n) \in \underset{\sim}{R}$$

and therefore $M \subset H(B)$. Given $b \in B$, there exists a sequence
$\{a_n\}$ in A with $b = \lim_{n \to \infty} a_n$. Moreover, we have $a_n = h_n + ik_n$
for some $h_n, k_n \in H(A) \subset H(B)$. By formula (1) in Lemma 5. 8,
we have $|h_n| \le e|h_n + ik_n|$, and therefore $\{h_n\}$ is a Cauchy
sequence in $H(A)$. Since $M \subset H(B)$, it follows that there is
$h \in H(B)$ such that $h = \lim_{n \to \infty} h_n$, and hence there is $k \in H(B)$
such that $k = \lim_{n \to \infty} k_n$. Therefore $b = h + ik$ and so
$B = H(B) + iH(B)$. By the Vidav-Palmer theorem B is a B*-
algebra with involution $(h + ik)^* = h - ik$. Since A is a star
subalgebra of B we conclude that A is a pre-B*-algebra.

Corollary 3. Let A be a complex Banach algebra with unit element, and let $|\,.\,|$ be a second algebra-norm on A such that $|1| = 1$ and $A = H(A, |\,.\,|) + iH(A, |\,.\,|)$. Then A is an A*-algebra.

Let A be a unital A*-algebra with Banach algebra norm $\|\,.\,\|$. The Hermitian elements with respect to $\|\,.\,\|$ may form a very small subspace. For example, let G be a group (with discrete topology) and let $A = \ell_1(G)$. It is well known that A is a unital A*-algebra. For the Banach algebra norm on A (i. e. the ℓ_1-norm) we may identify $D(A, 1)$ with the set of all complex functions f on G such that $f(e) = 1$ (e the identity of G) and $|f(x)| \leq 1$ ($x \in G$). It follows that the Hermitian elements of A (for the ℓ_1-norm) consist of the real multiples of the unit element.

It would be of interest to classify unital A*-algebras according to the size of H(A) for equivalent Banach algebra norms on A.

We next consider some results for algebras that do not have a unit element (or possibly have a unit element with norm greater than one). The definition of numerical range of an element a is as in 2.1 and is in fact the spatial numerical range (definition 9.1) of the left multiplication operator given by $T_a x = ax$. As before an element is Hermitian if it has real numerical range.

Theorem 4. Let A be a complex normed algebra such that $A = H(A) + iH(A)$, and the left regular representation is faithful. Then A is a pre-B*-algebra with the operator norm given by the left regular representation.

Proof. Let B be the subalgebra of the bounded operators on A generated by the identity operator and the image of A under the left regular representation $a \to T_a$. Then B is a complex unital normed algebra with the operator norm. Theorems 9. 4 and 2. 4 give

$$V(B, T_a) = \overline{co} \, V(A, a) \ ,$$

and hence $T_h \in H(B)$ whenever $h \in H(A)$. Since the unit element is always Hermitian and $A = H(A) + iH(A)$, it follows that $B = H(B) + iH(B)$. Therefore B is a pre-B*-algebra, by Theorem 2, and hence A is a pre-B*-algebra with norm defined by $|a| = |T_a|$.

Remarks. (1) If A is a pre-B*-algebra, it is easy to show that $A = H(A) + iH(A)$ and that the left regular representation is faithful. Thus, Theorem 4 gives a characterization of pre-B*-algebras.

(2) It is easy to construct examples of complex normed algebras A for which $A = H(A) + iH(A)$ while the left regular representation is not faithful, e. g. $A = \underset{\sim}{C}x$ where $x^2 = 0$.

(3) Let A satisfy the conditions of Theorem 4, and let A be complete. Then A is an A*-algebra. Moreover, if $\| . \|$ is the Banach algebra-norm on A, then A is a B*-algebra if and only if

$$\|a\| = \sup \{ \|ax\| : x \in A, \ \|x\| \leq 1 \} \qquad (a \in A).$$

In particular, if A has an approximate unit with norm ≤ 1, then A is a B*-algebra. Therefore, for complex Banach algebras with an approximate unit, the condition $A = H(A) + iH(A)$ is a characterization of B*-algebras. This last remark has also been observed independently by J. G. Pickford and T. W. Palmer.

(4) Let A be a semi-simple H*-algebra (see Rickart [57] page 272), so that the left regular representation $a \to T_a$ is a star isomorphism from A to the bounded operators on the Hilbert space A. With respect to the given norm, $V(a)$ is then the Hilbert space numerical range of T_a. Let $a^* = a$. Then $T_a^* = T_a$ and so $V(a) = V(T_a) \subset \mathbb{R}$. Let $\hat{A} = \{T_x : x \in A\}$. Then $V(\hat{A}, T_a) \subset V(B(A), T_a) = \overline{co} \, V(T_a)$, by Theorem 9.4. Thus the self-adjoint elements of A form the Hermitian elements of A with respect to the given norm $\|a\|$ and the operator norm $|a| = |T_a|$. In the case when A consists of the Hilbert-Schmidt operators on an infinite dimensional Hilbert space (see Rickart [57] page 285), the two norms $\| \cdot \|$, $| \cdot |$ are not equivalent.

(5) As a final example let G be a non-discrete locally compact group and let $A = L_1(G)$ so that A is an A*-algebra without unit element. We leave as an exercise the fact that $H(L_1(G)) = \{0\}$ with respect to the L_1-norm.

Our final application of the Vidav-Palmer theorem concerns the existence of maximal B*-algebras in an arbitrary complex unital Banach algebra A. Recall from §6 that we write $J(A)$ for the complex Banach space $H(A) + iH(A)$, and that $J(A)$ need not be an algebra. The theorem below has also been observed independently by A. M. Sinclair, and E. Torrance.[†]

---

† E. Torrance, Notices Amer. Math. Soc. (1969), 778.

Theorem 5. Let A be a complex unital Banach algebra such that $J(A)$ is an algebra. Then $J(A)$ is a B*-algebra, and if C is a subalgebra of A containing 1 that is a B*-algebra for some involution, then $C \subset J(A)$.

Proof. $J(A)$ is a complex unital Banach algebra and, by Theorem 2. 4, $J(A) = H(J(A)) + iH(J(A))$. Therefore $J(A)$ is a B*-algebra by the Vidav-Palmer theorem. Now let C be a subalgebra of A that is a B*-algebra for some involution. If h is self-adjoint in C, then it is Hermitian in C and hence in A. Therefore $C \subset J(A)$, as required.

By way of illustration let A be the Banach algebra of all bounded linear operators on the Banach space ℓ_1, i. e. $A = B(\ell_1)$. It is easily verified that $H(A)$ consists of the operators of the form

$$T\{x_n\} = \{\lambda_n x_n\}$$

where $\{\lambda_n\}$ is a bounded real sequence. Therefore $J(A)$ is an algebra and may be identified with the B*-algebra of all bounded complex sequences. The same result obtains if ℓ_1 is replaced by ℓ_p for $1 < p \leq \infty$, $p \neq 2$. If ℓ_1 is replaced by any complex Banach space X, then the size of $H(A)$ clearly gives some indication as to the presence of Hilbert space structure on X. More precisely, we have the following theorem.

Theorem 6. Let X be a complex Banach space and let $A = B(X)$. Then $A = H(A) + iH(A)$ if and only if X is a Hilbert space.

Proof. If X is a Hilbert space, A is a C*-algebra and therefore $A = H(A) + iH(A)$. Conversely, if $A = H(A) + iH(A)$, then A is a B*-algebra by the Vidav-Palmer theorem. By Rickart [57] Corollary (4.10.8), there is an inner product $(,)$ on X such that T* is the Hilbert space adjoint of T (with respect to the inner product) for each $T \in A$. Moreover, it follows from the argument of Rickart [57] Theorem (4.10.6) that $\|x\|^2 = (x, x)$ for each $x \in X$ so that X is already a Hilbert space.

As a final example let G be a locally compact group with identity element e, and let $A = M(G)$, the convolution measure algebra. Then A is a complex unital Banach algebra in which the unit element is the point mass measure at e. Given $f \in C_0(G)$ with $f(e) = \|f\| = 1$, it follows that

$$\int f \, d\mu \in V(A, \mu) \qquad (\mu \in A) \ .$$

A standard measure-theoretic argument may now be used to show that $\mu \in H(A)$ if and only if $\mu = \alpha 1$ for some $\alpha \in \underset{\sim}{R}$. It follows that $\underset{\sim}{C} 1$ is the only subalgebra of $M(G)$ containing 1 which is a B*-algebra.

3. Operators

Given a bounded linear operator T on a normed linear space X, we may regard T as an element of the unital normed algebra $B(X)$ of all bounded linear operators on X, and so have available the numerical range $V(B(X), T)$ and the results of chapters 1 and 2. However, a more natural numerical range of T is also available, defined directly in terms of the space X and its dual space, without intervention of the algebra $B(X)$. We denote this 'spatial' numerical range by $V(T)$.

In §9 we compare $V(T)$ with $V(B(X), T)$ and with the numerical range $W(T)$ corresponding to a semi-inner-product on X. §10 is concerned with spectral properties of $V(T)$. The principal result is a theorem of Williams that gives

$$Sp(T) \subset V(T)^-$$

when X is a complex Banach space. Theorems of Lumer and of Nirschl and Schneider give interesting spectral properties of boundary points of $V(T)$.

In general, $V(T)$ is not convex and it is therefore of interest to study its topological or geometrical properties. In §11 we show that $V(T)$ is connected, and that, except when $X \cong \mathbb{R}$, this result holds, without assumption of linearity, for any continuous mapping T of the unit sphere of X into X.

9. THE SPATIAL NUMERICAL RANGE

Let X denote a normed linear space over F, $S(X)$ its unit sphere $\{x \in X : \|x\| = 1\}$, and X' its dual space. Let B denote the normed algebra $B(X)$ of all bounded linear operators on X (i. e. bounded linear mappings of X into X) with the operator norm $|\,.\,|$, given by

$$|T| = \sup\{\|Tx\| : x \in X, \|x\| \le 1\}.$$

Let I denote the identity operator on X.

We denote by Π or $\Pi(X)$ the subset of the Cartesian product $X \times X'$ defined by

$$\Pi = \Pi(X) = \{(x, f) : x \in S(X), f \in S(X'), f(x) = 1\}.$$

Definition 1. The spatial numerical range $V(T)$ of an element T of B is defined by

$$V(T) = \{f(Tx) : (x, f) \in \Pi\}.$$

Note that if $D(X, x) = \{f \in X' : \|f\| = f(x) = 1\}$, then

$$V(T) = \cup \{\{f(Tx) : f \in D(X, x)\} : x \in S(X)\}.$$

It is almost obvious that

$$V(T) \subset V(B, T). \tag{1}$$

For, given $(x, f) \in \Pi$, define F on B by

$$F(S) = f(Sx) \qquad (S \in B) \ .$$

Then $F \in D(B, I)$, and so $f(Tx) = F(T) \in V(B, T)$.

Lemma 2. Let Γ be a subset of Π such that its natural projection $\pi_1(\Gamma) = \{x : (x, f) \in \Gamma$ for some $f\}$ is dense in $S(X)$. Then, for each $T \in B$,

$$\inf \{\tfrac{1}{\alpha}(\,|I + \alpha T\,| - 1) : \alpha > 0\,\} = \sup \{\mathrm{Re}\, f(Tx) : (x, f) \in \Gamma \, \} \ .$$

Proof. Let $\mu = \sup \{\mathrm{Re}\, f(Tx) : (x, f) \in \Gamma \}$. By (1) and Theorem 2.5, we have

$$\mu \leq \inf \{\tfrac{1}{\alpha}(\,|I + \alpha T\,| - 1) : \alpha > 0\} \ . \tag{2}$$

The case $T = 0$ is obvious; so assume that $T \neq 0$. Let $0 < \alpha < |T|^{-1}$, $\epsilon > 0$, $x \in S(X)$. Since $\pi_1(\Gamma)$ is dense in $S(X)$, there exists $(y, g) \in \Gamma$ such that $\|x - y\| < \epsilon$. We have

$$\mathrm{Re}\, g(Ty) \leq \mu \leq |T| \ ,$$

and so

$$\| (I - \alpha T)y \| \geq \mathrm{Re}\, g((I - \alpha T)y) = 1 - \alpha\, \mathrm{Re}\, g(Ty) \geq 1 - \alpha\mu > 0 \ .$$

Therefore

$$\| (I - \alpha T)x \| \geq 1 - \alpha\mu - |I - \alpha T|\, \epsilon \ .$$

Since ϵ is arbitrary, this gives $\|(I - \alpha T)x\| \geq 1 - \alpha\mu$, and therefore

$$\|(I - \alpha T)x\| \geq (1 - \alpha\mu)\|x\| \qquad (x \in X).$$

If we replace x by $(I + \alpha T)x$, this gives

$$\|(I + \alpha T)x\| \leq (1 - \alpha\mu)^{-1} \|(I - \alpha^2 T^2)x\| \qquad (x \in X),$$

and so

$$|I + \alpha T| \leq \frac{1 + \alpha^2 |T^2|}{1 - \alpha\mu} \quad .$$

Therefore

$$\frac{1}{\alpha} \{|I + \alpha T| - 1\} \leq \frac{\mu + \alpha |T^2|}{1 - \alpha\mu} \quad ,$$

and this with (2) completes the proof.

Theorem 3. Let Γ be a subset of Π such that its natural projection $\pi_1(\Gamma)$ is dense in $S(X)$. Then, for each $T \in B$,

$$\overline{co} \{f(Tx) : (x, f) \in \Gamma\} = V(B, T).$$

Here $\overline{co}\,E$ denotes the <u>closed convex hull</u> of a set E, i.e. the intersection of all closed convex sets containing E.

Proof. By Lemma 2 and Theorem 2.5, we have

$$\sup\{\operatorname{Re} f(Tx) : (x, f) \in \Gamma\} = \sup\{\operatorname{Re} \lambda : \lambda \in V(B, T)\}.$$

The proof is easily completed by replacing T by appropriate scalar multiples of T and using the fact that $V(B, T)$ is a closed convex set.

Theorem 4. For each $T \in B$, we have

(i) $\quad \overline{co} \, V(T) = V(B, T)$,

(ii) $\quad \sup \{ |\lambda| : \lambda \in V(T) \} = v(T) = \sup \{ |\lambda| : \lambda \in V(B, T) \}$.

Proof. Clear.

Theorem 5. For each $U \in B(X')$, we have

$$\overline{co} \, \{(Uf)(x) : (x, f) \in \Pi(X) \} = \overline{co} \, V(U) = V(B(X'), U) .$$

Proof. Given $x \in X$, let $\hat{x}$ be the element of X'' given by

$$\hat{x}(f) = f(x) \qquad (f \in X') .$$

By a theorem of Bishop and Phelps [9], the elements of X' that attain their bounds on the unit ball are dense in X'. Therefore if Γ is the subset of $\Pi(X')$ given by

$$\Gamma = \{(f, \hat{x}) : (x, f) \in \Pi(X) \}$$

then its projection $\pi_1(\Gamma)$ is dense in $S(X')$. By Theorem 3,

$$\overline{co} \, \{(Uf)(x) : (x, f) \in \Pi(X) \} = \overline{co} \, \{\hat{x}(Uf) : (f, \hat{x}) \in \Gamma \} = V(B(X'), U).$$

Corollary 6. Let T^* denote the adjoint of $T \in B(X)$. Then

(i) $V(T) \subset V(T^*)$,

(ii) $\overline{co} \, V(T) = \overline{co} \, V(T^*)$,

(iii) $v(T) = v(T^*)$.

Proof. Immediate.

Definition 7. A <u>semi-inner-product</u> (s. i. p.) on X is a mapping $[,]$ of $X \times X$ into $\underset{\sim}{F}$ such that:

(i) the mapping $x \to [x, y]$ is linear on X for each fixed $y \in X$;

(ii) $[x, x] > 0$ if $x \neq 0$;

(iii) $|[x, y]|^2 \leq [x, x][y, y]$ $\qquad (x, y \in X)$.

It is proved in Lumer ([40] Theorem 2) that the mapping $x \to [x, x]^{\frac{1}{2}}$ is a norm on X. We say that a s. i. p. $[,]$ <u>determines the norm</u> of X, if

$$\|x\| = [x, x]^{\frac{1}{2}} \qquad (x \in X) \, ,$$

where $\| . \|$ is the given norm on X.

Given a s. i. p. $[,]$ and $T \in B$, the <u>numerical range</u> $W(T)$ corresponding to $[,]$ is defined by

$$W(T) = \{[Tx, x] : [x, x] = 1 \} \, .$$

Theorem 8. Let $[,]$ be a s. i. p. that determines the norm of X, and $W(T)$ the corresponding numerical range. Then for $T \in B$,

(i) $W(T) \subset V(T)$,

(ii) $\overline{co}\, W(T) = V(B, T)$,

(iii) $\sup \{ \operatorname{Re} \lambda : \lambda \in W(T) \} = \inf \{ \frac{1}{\alpha} (|I + \alpha T| - 1) : \alpha > 0 \}$,

(iv) $\sup \{ |\lambda| : \lambda \in W(T) \} = v(T)$.

Proof. Given $y \in S(X)$, let f_y be defined on X by

$$f_y(x) = [x, y] .$$

Then $f_y \in X'$, $\|f_y\| \le 1$, and $f_y(y) = [y, y] = \|y\|^2 = 1$. Therefore $(y, f_y) \in \Pi$. Let $\Gamma = \{ (y, f_y) : y \in S(X) \}$. Then $\pi_1(\Gamma) = S(X)$, and so Theorem 3 is applicable.

Remarks. (1) Let X be a normed linear space. We know, by the Hahn-Banach theorem, that $D(X, y)$ is a non-empty subset of $S(X')$ for each $y \in S(X)$. Let ψ be a mapping of $S(X)$ into $S(X')$ such that

$$\psi(y) \in D(X, y) \qquad (y \in S(X)) ,$$

and let $[,]$ be defined on $X \times X$ by

$$[x, y] = \begin{cases} 0 & (y = 0) \\ \|y\|\, \psi\, (\|y\|^{-1} y)(x) & (y \ne 0) . \end{cases}$$

It is easy to check that $[,]$ is a s. i. p. on **X** that determines the norm of **X**. Thus there always exists such a s. i. p., and in general many corresponding to different choices of ψ belonging to the Cartesian product,

$$X \{D(X, y): y \in S(X) \} .$$

It follows at once that $V(T)$ is the union of all the numerical ranges $W(T)$ corresponding to the semi-inner-products that determine the norm of **X**.

(2)   Lemma 2, which is the crucial result in this section is essentially due to Lumer ([40] Lemma 12).

(3)   Let the numerical index of **X** be the real number $n(X)$ defined by

$$n(X) = \inf \{v(T) : T \in B(X), \ |T| = 1 \} .$$

If **X** is a complex normed linear space, Theorem 4.1 shows that

$$\frac{1}{e} \le n(X) \le 1 .$$

It has long been known that, for a complex Hilbert space **X** of dimension greater than one, $n(X) = \frac{1}{2}$ (Halmos [30] page 114). Glickfeld [28] gives an example of a norm on $\underset{\sim}{C}^2$ for which $n(\underset{\sim}{C}^2) = \frac{1}{e}$. Duncan, McGregor, Pryce and White [21] prove that if **E** is a compact Hausdorff space, then $n(C(E)) = 1$; and hence, by Corollary 6, $n(X) = 1$ for all complex **L** and **M** spaces. They also prove that for every real number t with $\frac{1}{e} \le t \le 1$, there exists a norm on $\underset{\sim}{C}^2$ which gives $n(\underset{\sim}{C}^2) = t$; and for every real number t with $0 \le t \le 1$, there exists a norm on $\underset{\sim}{R}^2$ which gives $n(\underset{\sim}{R}^2) = t$.

(4) So far as we know, the numerical index has not yet been computed for other classical Banach spaces, L_p $(1 < p < \infty,\ p \ne 2)$ and the disc algebra $A(\Delta)$, for example. [†]

(5) Let X be a real or complex finite dimensional normed space. C. M. McGregor (private communication) has shown that $n(X) = 1$ if and only if $|f(x)| = 1$ whenever x is an extreme point of the unit ball of X and f is an extreme point of the unit ball of X'.

10. SPECTRAL PROPERTIES

Throughout this section, let X be a complex Banach space, let $T \in B(X)$, and let $Sp(T)$ denote $Sp(B(X), T)$.

By Theorem 2. 6, we have

$$Sp(T) \subset V(B(X), T) = \overline{co}\ V(T) .$$

However, the following stronger statement holds.

Theorem 1.   (Williams [74].) $Sp(T) \subset V(T)^-$.

Proof.   Let $\lambda \in \underset{\sim}{C}$ and $\inf \{|\lambda - \zeta| : \zeta \in V(T)\} = \epsilon > 0$. Then, for each $(x, f) \in \Pi$, we have

---

[†] $n(A(\Delta)) = 1$ (M. J. Crabb).

$$\|(\lambda I - T)x\| \geq |f((\lambda I - T)x)| = |\lambda - f(Tx)| \geq \epsilon , \qquad (1)$$

and similarly

$$\|(\lambda I - T)^*f\| \geq \epsilon . \qquad (2)$$

Given $g \in S(X')$ and $\delta > 0$, the theorem of Bishop and Phelps [9] implies the existence of $(x, f) \in \Pi$ with $\|f - g\| < \delta$. Then, by (2),

$$\|(\lambda I - T)^*g\| \geq \|(\lambda I - T)^*f\| - |\lambda I - T|\delta \geq \epsilon - |\lambda I - T|\delta ;$$

and, since δ and g are arbitrary, this gives

$$\|(\lambda I - T)^*g\| \geq \epsilon \|g\| \qquad (g \in X') . \qquad (3)$$

From (1),

$$\|(\lambda I - T)x\| \geq \epsilon \|x\| \qquad (x \in X) . \qquad (4)$$

The inequality (4) shows that $\lambda I - T$ is a one-to-one mapping of X onto a closed subspace Y, and (3) shows that Y is dense in X. Therefore $Y = X$, and Banach's isomorphism theorem shows that $\lambda I - T$ is an invertible element of $B(X)$; and so $\lambda \notin Sp(T)$.

Remark. It is of course trivial that every eigenvalue of T is actually in $V(T)$.

Theorem 2. (Williams [74].) Let $S, T \in B(X)$, let $0 \notin V(T)^-$, and let $E = \{\lambda \mu^{-1} : \lambda \in V(S)^-, \mu \in V(T)^- \}$.

89

Then

$$Sp(T^{-1}S) \subset E .$$

Proof. Let ζ be a complex number not belonging to E. Then there exists $\delta > 0$ such that

$$|\zeta\mu - \lambda| \geq \delta \qquad (\lambda \in V(S)^-, \quad \mu \in V(T)^-) .$$

Given $(x, f) \in \Pi$, we have

$$\|(\zeta T - S)x\| \geq |f((\zeta T - S)x)| = |\zeta f(Tx) - f(Sx)| \geq \delta ,$$

since $f(Tx) \in V(T)$ and $f(Sx) \in V(S)$. Similarly

$$\|(\zeta T - S)^*f\| \geq \delta .$$

Applying the theorem of Bishop and Phelps as in the proof of Theorem 1, we conclude that $\zeta T - S$ is invertible. Since $0 \notin V(T)^-$ and $Sp(T) \subset V(T)^-$, T is invertible. Therefore $\zeta I - T^{-1}S$ is invertible, i. e.

$$\zeta \notin Sp(T^{-1}S) .$$

The next theorem is a slightly strengthened form of Theorem 2.10. For its proof we need the following lemma.

Lemma 3. Let S be a bounded semi-group in $B(X)$. Then there exists a norm $\|.\|_1$ on X, equivalent to the given norm, such that

$$\|Tx\|_1 \leq \|x\|_1 \quad (x \in X, \ T \in S).$$

Proof. We have $|T| \leq M$ $(T \in S)$, and we may assume that $I \in S$. Then $\|\,.\,\|_1$, defined by

$$\|x\|_1 = \sup\{\|Tx\| : T \in S\} \quad (x \in X),$$

has the required properties.

Theorem 4. $\text{co } Sp(T) = \cap \overline{\text{co}}\ V(T)$, where the intersection is taken over all numerical ranges $V(T)$ corresponding to norms on X equivalent to the given norm.

Proof. By Theorem 1, $\text{co } Sp(T) \subset \cap \overline{\text{co}}\ V(T)$. The rest of the proof is similar to the proof of Theorem 2.10, but with Lemma 2.7 replaced by Lemma 3.

Remark. Since $\overline{\text{co}}\ V(T) = V(B(X), T)$, Theorem 4 differs from Theorem 2.10 in that we need only take those algebra-norms on $B(X)$ that are operator norms corresponding to norms on X equivalent to the given norm.

See also Holmes [33], where it is proved that the spectral radius of a bounded linear operator T is the infimum of the operator-norms of T corresponding to equivalent norms on the underlying space.

Theorem 5. (Lumer [40].) For each s.i.p. that determines the norm of X, the corresponding numerical range $W(T)$ satisfies

$$\partial Sp(T) \subset W(T)^-,$$

where ∂ denotes 'the topological boundary of'.

Proof. Let $\lambda \in \partial \mathrm{Sp}(T)$. Then λ belongs to the approximate point spectrum of T; i.e. there exist $x_n \in S(X)$ such that $(\lambda I - T)x_n \to 0$ as $n \to \infty$. If $[,]$ is a s.i.p. that determines the norm of X, we have

$$\left| \lambda - [Tx_n, x_n] \right| = \left| [(\lambda I - T)x_n, x_n] \right| \leq \left\| (\lambda I - T)x_n \right\| ,$$

and so $\lambda \in W(T)^-$.

Theorem 6. (Lumer [40].) Let X be uniformly convex. Then

$$\{\lambda \in V(T)^- : |\lambda| = |T| \} \subset \partial \mathrm{Sp}(T) .$$

Corollary 7. If X is uniformly convex and $v(T) = |T|$, then $\rho(T) = |T|$.

Proof. Let $\lambda \in V(T)^-$ and $|\lambda| = |T|$. We may assume that $\lambda \neq 0$, for otherwise $T = 0$ and $\mathrm{Sp}(T) = \{0\}$. Since we may replace T by $\frac{1}{\lambda}T$, there is no loss of generality in assuming that $\lambda = 1$. Then there exist $(x_n, f_n) \in \Pi$ such that $f_n(Tx_n) \to 1$ and therefore $f_n(\frac{1}{2}(x_n + Tx_n)) \to 1$ as $n \to \infty$. Since

$$1 \geq \left\| \tfrac{1}{2}(x_n + Tx_n) \right\| \geq \left| f_n(\tfrac{1}{2}(x_n + Tx_n)) \right| ,$$

it follows that $\left\| \tfrac{1}{2}(x_n + Tx_n) \right\| \to 1$. But X is uniformly convex (Clarkson [16], uniformly rotund, Day [19]), and therefore $x_n - Tx_n \to 0$. We have $\|x_n\| = 1$ and $(I - T)x_n \to 0$, which show that $1 \in Sp(T)$. Since $|T| = 1$, this gives $1 \in \partial Sp(T)$. The corollary is immediate.

In the following theorem a weaker condition is imposed on X and a stronger condition on λ and we have the stronger conclusion that λ is an eigenvalue. A normed space is said to be strictly convex (Clarkson [16], rotund,Day [19]) if and only if x and y are linearly dependent whenever

$$\|x + y\| = \|x\| + \|y\| .$$

Theorem 8.   Let X be strictly convex, let $\lambda \in V(T)$, and let $|\lambda| = |T|$. Then λ is an eigenvalue of T.

Proof.   We may assume that $\lambda = 1 = |T|$. Then there exists $(x, f) \in \Pi$ such that $f(Tx) = 1$. Therefore

$$2 \geq \|x\| + \|Tx\| \geq \|x + Tx\| \geq |f(x + Tx)| = 2 .$$

This proves that $\|x + Tx\| = \|x\| + \|Tx\|$, and so x and Tx are linearly dependent. It is now easy to show that $Tx = x$.

Lemma 9.   Let $V(T) \subset \{\zeta \in \underset{\sim}{C} : Re\ \zeta \leq 0\}$. Then

$$\|T^2 x\| \geq \delta(\|\delta x + Tx\| - \delta\|x\|) \qquad (x \in X, \quad \delta \geq 0) .$$

Proof.   Let $u = \delta x + Tx$. If $u = 0$, the required inequality is obvious. Assume that $u \neq 0$, take $v = \dfrac{1}{\|u\|} u$, and choose $f \in D(X, v)$. Then

$$\|u\| = \|u\| f(v) = f(u) = \delta f(x) + f(Tx) \, ,$$

and so

$$\|u\| f(Tv) = f(Tu) = \delta f(Tx) + f(T^2x) = \delta \|u\| - \delta^2 f(x) + f(T^2x) \, .$$

Since $\operatorname{Re} f(Tv) \le 0$, this gives

$$\|T^2x\| \ge \left| f(T^2x) \right| \ge -\operatorname{Re} f(T^2x) \ge \delta \|u\| - \delta^2 \operatorname{Re} f(x)$$
$$\ge \delta \|u\| - \delta^2 \|x\|$$
$$\ge \delta(\|\delta x + Tx\| - \delta \|x\|) \, .$$

The following theorem shows that eigenvalues of T that lie on the boundary of $\overline{co}\, V(T)$ have index (ascent) 1.

Theorem 10. (Nirschl and Schneider [47].) Let $\lambda \in \partial \overline{co}\, V(T)$. If $(\lambda I - T)^2 x = 0$, then $(\lambda I - T)x = 0$.

Proof. We may assume that $\lambda = 0$ and that $V(T) \subset \{ \zeta \in \underset{\sim}{C} : \operatorname{Re} \zeta \le 0 \}$. By Lemma 9, if $T^2x = 0$, we have $\|\delta x + Tx\| - \delta \|x\| \le 0$ $(\delta > 0)$, and so $Tx = 0$.

Remark. M. J. Crabb has proved that eigenvalues of index greater than one are interior points of $V(T)$. In fact, if $\|u\| = 1$ and $T^2u = 0 \ne Tu$, then $V(T)$ contains the disc $\{ \zeta \in \underset{\sim}{C} : |\zeta| < \frac{1}{6} \|Tu\| \}$. The details will be published elsewhere.

Corollary 11. Let T be a Hermitian operator, and let $\lambda \in \operatorname{Sp}(T)$. If $(\lambda I - T)^2 x = 0$, then $(\lambda I - T)x = 0$.

Proof. Since $\mathrm{Sp}(T) \subset V(T)^{-} \subset \underset{\sim}{\mathbf{R}}$, $\lambda \in \partial \overline{\mathrm{co}}\, V(T)$.

Corollary 12. Let $\Delta = \{\zeta \in \underset{\sim}{\mathbf{C}} : |\zeta| \leq 1\}$, let T be invertible, and suppose that $V(T) \subset \Delta$ and $V(T^{-1}) \subset \Delta$. If $\lambda \in \mathrm{Sp}(T)$ and $(\lambda I - T)^2 x = 0$, then $(\lambda I - T)x = 0$.

Proof. We have $\mathrm{Sp}(T) \subset \Delta$ and $\mathrm{Sp}(T^{-1}) \subset \Delta$. Therefore $\mathrm{Sp}(T) \subset \{\zeta \in \underset{\sim}{\mathbf{C}} : |\zeta| = 1\}$, and all points of $\mathrm{Sp}(T)$ are boundary points of $\overline{\mathrm{co}}\, V(T)$.

Theorem 13. (i) If $V(T) \subset \{\zeta : \mathrm{Re}\ \zeta \leq 0\}$, then

$$\|T^2 x\| \geq \frac{1}{8} \|Tx\|^2 \qquad (x \in S(X)) .$$

(ii) If T is Hermitian, then

$$\|T^2 x\| \geq \frac{1}{4} \|Tx\|^2 \qquad (x \in S(X)) .$$

Proof. (i) By Lemma 9, if $V(T) \subset \{\zeta : \mathrm{Re}\ \zeta \leq 0\}$, we have

$$\|T^2 x\| \geq \delta\, (\|Tx\| - 2\delta) \qquad (x \in S(X),\ \delta \geq 0) .$$

Take $\delta = \frac{1}{4} \|Tx\|$.

(ii) Suppose $V(T) \subset \{\zeta : \mathrm{Re}\ \zeta = 0\}$. Then Lemma 9 is applicable to T and to $-T$, and we have

$$\|T^2 x\| \geq \delta(\|Tx + \delta x\| - \delta \|x\|) ,$$
$$\|T^2 x\| \geq \delta(\|Tx - \delta x\| - \delta \|x\|) .$$

Since $\|Tx + \delta x\| + \|Tx - \delta x\| \geq 2\|Tx\|$, this gives

$$\|T^2x\| \geq \delta(\|Tx\| - \delta\|x\|).$$

Take $x \in S(X)$ and $\delta = \frac{1}{2}\|Tx\|$.

Finally, given a Hermitian operator T, we have $V(iT) \subset \{\zeta : \text{Re } \zeta = 0\}$.

We conclude this section with two applications of the Nirschl-Schneider result (Theorem 10 above). The first application is a simple proof of a theorem on stochastic matrices (see Gantmacher [23] pages 84-87). Recall that an $n \times n$ matrix $A = (\alpha_{ij})$ is said to be stochastic if

$$\alpha_{ij} \geq 0 \ (i, j = 1, \ldots, n); \ \sum_{j=1}^{n} \alpha_{ij} = 1 \ (i = 1, \ldots, n).$$

Theorem 14. Let A be a stochastic matrix. Then $\rho(A) = 1$ and all eigenvalues λ with $|\lambda| = 1$ have index one.

Proof. Let $A = (\alpha_{ij})$ be $n \times n$, and let $X = \underset{\sim}{C}^n$ with the supremum norm. If $e = (1, 1, \ldots, 1)$, then $Ae = e$. Therefore $1 \in \text{Sp}(A)$. Given $x = (\xi_j)$, we have

$$|(Ax)_i| = |\sum_{j=1}^{n} \alpha_{ij}\,\xi_j| \leq \sum_{j=1}^{n} \alpha_{ij}\,\|x\| = \|x\|$$

and therefore

$$\|Ax\| \leq \|x\| \qquad (x \in X).$$

This gives $1 \leq \rho(A) \leq |A| \leq 1$, and so

$$\rho(A) = v(A) = |A| = 1 \, .$$

Therefore all eigenvalues λ with $|\lambda| = 1$ are frontier points of co $V(A)$ and Theorem 10 completes the proof.

Our final application of the Nirschl-Schneider result is a sharpening of power inequalities on the numerical radius of special operators (see Theorem 4. 6). The theorem below is due to M. J. Crabb [17].

Theorem 15. Let X be finite dimensional and let $T \in B(X)$ with $v(T) = 1$. Then $\{|T^n|\}$ is a bounded sequence.

Proof. Let $\lambda_1, \ldots, \lambda_r$ be the eigenvalues of T of modulus one. Then, by Theorem 10, each such eigenvalue has index one. It follows that the Jordan canonical form of T is given by

$$S = \begin{pmatrix} \Lambda & 0 \\ 0 & U \end{pmatrix}$$

where $\Lambda = \text{diag}\{\lambda_1, \ldots, \lambda_r\}$ and $\lim_{n \to \infty} |U^n| = 0$. Moreover there is an invertible operator $P \in B(X)$ such that $S = P^{-1}TP$. Therefore $T^n = P S^n P^{-1}$ and it follows that $\{|T^n|\}$ is a bounded sequence.

Remark. The above theorem is readily extended to meromorphic operators on an arbitrary complex Banach space.

97

11. GEOMETRIC AND TOPOLOGICAL PROPERTIES OF V(T)

An early theorem in the theory of the numerical range asserts that the numerical range $W(T)$ of a bounded linear operator on a pre-Hilbert space is convex, and, for every bounded linear operator T on a normed linear space X, we know that $V(B(X), T)$ is convex. We show that the spatial numerical range $V(T)$ is not in general convex, even for linear operators on $\underset{\sim}{C}^2$, but that it is always connected. With one trivial exception this last result holds also for non-linear continuous operators.

Example 1. Let $1 < p < \infty$ and $q = p/(p-1)$ so that $\frac{1}{p} + \frac{1}{q} = 1$; and let $X_p = \underset{\sim}{C}^2$ with the ℓ_p norm,

$$\| (z, w) \|_p = \{ |z|^p + |w|^p \}^{\frac{1}{p}} \, .$$

Let T be the linear operator on X_p given by

$$T(z, w) = (iz + w, -(z + iw)) \, ,$$

and $V_p(T)$ denote its spatial numerical range.

Given $(z, w) \in S(X_p)$, we know from consideration of the case of equality in Hölder's inequality that there is a unique vector $(\lambda, \mu) \in S(X_q)$ with $\lambda z + \mu w = 1$, namely

$$(\lambda, \mu) = (z^* |z|^{p-2}, \ w^* |w|^{p-2}) \, .$$

Therefore

$$V_p(T) = \{i|z|^p - i|w|^p + wz^*|z|^{p-2} - w^*z|w|^{p-2} :$$

$$|z|^p + |w|^p = 1\} . \tag{1}$$

Writing $z = r\exp(i\phi)$, $w = s\exp(i\psi)$, $\theta = \psi - \phi$, $r \geq 0$, $s \geq 0$, $\phi, \psi \in \underset{\sim}{R}$, we therefore have

$$V_p(T) = \{\cos\theta . \; rs(r^{p-2} - s^{p-2}) + i[r^p - s^p + \sin\theta .$$

$$rs(r^{p-2} + s^{p-2})] : r^p + s^p = 1\} .$$

It follows that

$$\alpha = \sup\{\operatorname{Re}\zeta : \zeta \in V_p(T)\} = \sup\{rs(r^{p-2} - s^{p-2}):$$

$$r^p + s^p = 1\} ,$$

and

$$\beta = \sup\{V_p(T) \cap \underset{\sim}{R}\} = \sup\{\cos\theta . \; rs(r^{p-2} - s^{p-2}):$$

$$r^p + s^p = 1, \; r^p - s^p + \sin\theta . \; rs(r^{p-2} + s^{p-2}) = 0\} .$$

The suprema α and β are attained with say

$$\alpha = r_1 s_1(r_1^{p-2} - s_1^{p-2}), \quad \beta = \cos\theta \; r_2 s_2(r_2^{p-2} - s_2^{p-2}) ,$$

and we have $\alpha > 0$ unless $p = 2$. Plainly, if $p \neq 2$, we have $\beta < \alpha$ unless $\cos\theta = 1$. But since $r_2^p - s_2^p + \sin\theta . \; r_2 s_2(r_2^{p-2} + s_2^{p-2}) = 0$, we have $r_2 = s_2$ if $\cos\theta = 1$, giving $\beta = 0$. Thus $\alpha > \beta$ unless $p = 2$.

From equation (1), we see that $(V_p(T))^* = V_p(T)$. Thus α is attained at points above and below the real axis, and we conclude that $V_p(T)$ is not convex unless $p = 2$.

Remark. That $V(T)$ need not be convex was pointed out by Bauer [3], and essentially the above operator T on $\underset{\sim}{C}^2$ with the ℓ_∞ norm was given by Nirschl and Schneider [47] as an example of a non-convex $V(T)$. Likewise $V(T)$ is not convex when T is regarded as an operator on $\underset{\sim}{C}^2$ with the ℓ_1-norm.

It is tempting to conjecture that the convexity of $V(T)$ for every $T \in B(X)$ may characterize pre-Hilbert spaces X, see Zenger [78].

We turn now to topological properties of $V(T)$. Since the numerical range $W(T)$ of a bounded operator T on a Hilbert space need not be a closed set, the same is true of $V(T)$. In what follows we show that $V(T)$ is connected.

We use the notation of §9.

Definition 2. The norm $\times$ weak* topology is the product topology on $X \times X'$ given by the norm topology on X and the weak* topology on X'.

Lemma 3. Let π_1 denote the natural projection $\pi_1(x, f) = x$ of $X \times X'$ onto X, and let E be a subset of Π that is relatively closed in Π with respect to the norm $\times$ weak* topology. Then $\pi_1 E$ is a (norm) closed subset of X.

Proof. Let $x_n \in \pi_1 E$ and $x_n \to x \in X$. Then there exist elements f_n of $S(X')$ such that $(x_n, f_n) \in E$. By the weak* compactness of the closed unit ball in X', there exists a weak* cluster point f of the sequence $\{f_n\}$ with $\|f\| \leq 1$. We have

$$f(x) = (f - f_n)(x) + f_n(x - x_n) + f_n(x_n) ,$$

and so

$$\left| f(x) - 1 \right| \leq \left| (f - f_n)(x) \right| + \|x - x_n\| .$$

Since $\left| (f - f_n)(x) \right|$ is small for some arbitrarily large n, and $\|x - x_n\|$ is small for all sufficiently large n, it follows that $f(x) = 1$, and therefore $(x, f) \in \Pi$. But E is relatively closed in Π for the norm $\times$ weak* topology, and so $(x, f) \in E$ and $x \in \pi_1 E$.

Theorem 4. (Bonsall, Cain, and Schneider [12].) Π is a connected subset of $X \times X'$ with the norm $\times$ weak* topology unless X has dimension one over $\underset{\sim}{R}$.

Proof. Suppose that $\Pi = A \cup B$ with A, B relatively closed in Π for the norm $\times$ weak* topology, and that $A \cap B = \emptyset$. Then $\pi_1 A$ and $\pi_1 B$ are closed subsets of X, by Lemma 3, and

$$S(X) = \pi_1 A \cup \pi_1 B .$$

Suppose that $x \in \pi_1 A \cap \pi_1 B$. Then there exist $f, g \in X'$ such that $(x, f) \in A$ and $(x, g) \in B$. But then

$$(x, \alpha f + (1 - \alpha)g) \in \Pi \qquad (0 \leq \alpha \leq 1) ,$$

which is impossible since $A \cap B = \emptyset$. Therefore

$$\pi_1 A \cap \pi_1 B = \emptyset.$$

Suppose next that X does not have dimension one over $\underset{\sim}{R}$. If $x, y \in S(X)$ and $x + y \neq 0$, then

$$z_\alpha = \alpha x + (1 - \alpha)y \neq 0 \qquad (0 \leq \alpha \leq 1),$$

and so $\|z_\alpha\|^{-1} z_\alpha \in S(X)$ $(0 \leq \alpha \leq 1)$, which shows that x, y belong to the same component of $S(X)$. If $x, y \in S(X)$ and $x + y = 0$, we choose $z \in S(X)$ such that x, z are linearly independent. Then x, z belong to the same component, and y, z belong to the same component, and consequently x, y belong to the same component. Therefore $S(X)$ is connected.

Finally, let X have dimension one over $\underset{\sim}{R}$, and take $u \in X$ with $\|u\| = 1$. Then every $x \in X$ is of the form $x = \xi u$ with $|\xi| = \|x\|$, and, for every $f \in X'$, we have $f(x) = \xi f(u)$, $\|f\| = |f(u)|$. Let g be the functional given by

$$g(x) = \xi \qquad (x = \xi u \in X).$$

Then $S(X) = \{u, -u\}$, and Π has exactly two points (u, g) and $(-u, -g)$. Plainly, Π is disconnected in this case.

Corollary 5. $V(T)$ is connected.

Proof. We have

$$|f(Tx) - g(Ty)| \leq \|Tx - Ty\| + |(f - g)(Ty)| \quad ((x, f), (y, g) \in \Pi).$$

Therefore the mapping $(x, f) \to f(Tx)$ is a continuous mapping of Π

with the relative norm × weak* topology onto V(T). Therefore, by Theorem 4, V(T) is connected, except perhaps when X has dimension one over $\underset{\sim}{R}$.

We have seen that in this exceptional case Π has exactly two elements (u, g), (-u, -g), and so V(T) has exactly one point g(Tu).

Corollary 6. Let F be a continuous mapping of S(X) into X, and let V(F) = {f(F(x)) : (x, f) ∈ Π}. Then V(F) is connected, except perhaps when X has dimension one over $\underset{\sim}{R}$.

Proof. As for Corollary 5.

Example 7. In this example [,] is a semi-inner-product that determines the norm of X, T ∈ B(X), and W(T) is not connected.

Let $X = \underset{\sim}{F}^2$ with the sup norm $\|x\| = \max(|\xi_1|, |\xi_2|)$ $(x = (\xi_1, \xi_2) \in X)$, and let [x, y] be defined for $x = (\xi_1, \xi_2)$, $y = (\eta_1, \eta_2)$, by

$$[x, y] = \begin{cases} \xi_1 \eta_1^* & (\text{if } |\eta_1| = \|y\|) , \\ \xi_2 \eta_2^* & (\text{if } |\eta_1| < \|y\|) . \end{cases}$$

Then [,] is a semi-inner-product on X, $[x, x] = \|x\|^2$, and for the operator T given by

$$Tx = (\xi_1, 0) \quad (x = (\xi_1, \xi_2) \in X) ,$$

we have W(T) = {0, 1}. For if $\|x\| = 1$ and $|\xi_1| = 1$, then

$[Tx, x] = \xi_1 \xi_1^* = |\xi_1|^2 = 1$, and if $\|x\| = 1$ and $|\xi_1| < 1$, then $[Tx, x] = 0$. $\xi_2^* = 0$. Thus $W(T)$ is disconnected.

We do not know of any example in which $V(T)$ is not simply connected, i.e. in which $\underset{\sim}{C} \setminus V(T)$ is not connected. It is however easy to give an example of a non-linear mapping F for which $V(F)$ is not simply connected.

Example 8. Let $X = \underset{\sim}{C}$ with the usual norm, and let F be the constant valued mapping

$$F(x) = 1 \quad (x \in S(X)) .$$

Then $V(F)$ is the unit circle $\{\zeta \in \underset{\sim}{C} : |\zeta| = 1\}$.

4. Some Recent Developments

This chapter contains a miscellany of results. We discuss in §12 the numerical range in the second dual of a Banach algebra with the Arens multiplication. As an application of the Vidav-Palmer theorem we give a simple proof of the known result that the second dual of a complex B*-algebra is again a B*-algebra. In §13 we initiate the study of spectral states of a complex unital Banach algebra. A spectral state is a linear functional normalized at the identity element and dominated by the spectral radius. The existence of spectral states is related to commutativity in the algebra in that every commutator (element of the form ab - ba) is annihilated by every spectral state. We show in particular that there are no spectral states on B($\mathcal{H}$) where $\mathcal{H}$ is an infinite dimensional Hilbert space. The simplest examples of spectral states are given by multiplicative linear functionals and we show that these are the only "irreducible" spectral states. For finite-dimensional algebras we give a complete description of the spectral states in terms of normalized traces on matrix algebras.

The final section contains some remarks on isolated topics not covered in the main exposition, announcements of some recent results, and selected open problems on numerical range.

12. THE SECOND DUAL OF A BANACH ALGEBRA

The discussion of the second dual of a Banach algebra has been deferred, although it more appropriately belongs in chapter 1, because we need results from §9. Using a construction of the multiplication due to Arens [1], [2] the second dual space of a Banach algebra becomes a Banach algebra. We show that the Vidav-Palmer characterization of B*-algebras gives a straight-forward intrinsic proof that the second dual of a B*-algebra with the Arens multiplication is a B*-algebra. (See Civin and Yood [15], Sherman [60], Takeda [69] for a proof using representation theory, see also Tomita [71].) The proof also shows that the involution on the second dual is the natural one derived from the involution of the given B*-algebra.

Definition 1. Let A be a normed algebra. Given $f \in A'$, $x \in A$, let $<f, x>$ be the element of A' defined by

$$<f, x> (y) = f(xy) \qquad (y \in A) . \tag{1}$$

Given $F \in A''$, $f \in A'$, let $[F, f]$ be the element of A' defined by

$$[F, f] (x) = F(<f, x>) \qquad (x \in A) . \tag{2}$$

Finally, given $F, G \in A''$, let FG be the element of A'' defined by

$$(FG)(f) = F([G, f]) \qquad (f \in A') . \tag{3}$$

It is straightforward to verify that with this multiplication, the Arens multiplication, A" is a Banach algebra. Given a ∈ A, let $\hat{a}$ denote its canonical image in A", i. e.

$$\hat{a}(f) = f(a) \quad (f \in A') \, .$$

Then the canonical mapping a → $\hat{a}$ is an isometric isomorphism of A into A". Also, if A has a unit element 1, then

$$< f, 1 > = f \quad (f \in A') \, , \tag{4}$$

and $\hat{1}$ is a unit element for A".

A second multiplication F· G is constructed on A" by replacing equations (1), (2), (3), by

$$< f \,|\, x > \; (y) = f(yx) \quad (y \in A) \, ,$$

$$[F \,|\, f] \, (x) = F(< f \,|\, x >) \quad (x \in A) \, ,$$

$$(F \cdot G)(f) = F([G \,|\, f]) \quad (f \in A') \, .$$

This is the Arens multiplication on A" corresponding to the reversed multiplication x· y on A defined by

$$x \cdot y = yx \quad (x, y \in A) \, .$$

However, this second multiplication F· G on A" is not in general the reversed Arens multiplication, i. e. we do not in general have F· G = GF. The Arens multiplication is said to be regular if F· G = GF for all F, G in A". (Civin and Yood [15].)

Theorem 2. Let A be a unital normed algebra. Then A"
with the Arens multiplication is a unital Banach algebra, and
given $F \in A"$,

$$V(A", F) \; = \; \{F(f) : f \in D(A, 1)\}^{-} .$$

Corollary 3. $F \in A"$ is Hermitian if and only if
$F(f) \in \underset{\sim}{R}$ for every $f \in D(A, 1)$.

Proof. Let $F \in A"$, and let T denote the left regular
representation of F on A", i.e. the element T of B(A") such
that

$$TG = FG \quad (G \in A") .$$

By Theorem 9.5,

$$V(A", F) = V(T) = \overline{co} \; \{(TG)(g) : (g, G) \in \Pi(A')\}$$
$$= \overline{co} \; \{(FG)(g) : (g, G) \in \Pi(A')\}$$
$$= \overline{co} \; \{F([G, g]) : (g, G) \in \Pi(A')\} .$$

By (4), $[G, g] (1) = G(< g, 1 >) = G(g)$, and also $\|[G, g]\| \leq$
$\|G\| \; \|g\|$. Therefore $[G, g] \in D(A, 1)$ whenever $(g, G) \in \Pi(A')$,
and so

$$V(A", F) \subset \overline{co} \; \{F(f) : f \in D(A, 1)\} .$$

On the other hand, given $f \in D(A, 1)$, let Φ denote its
canonical image in A''', i.e.

$$\Phi(G) = G(f) \quad (G \in A") .$$

Then $\Phi(\hat{1}) = \hat{1}(f) = f(1) = 1$, and $\|\Phi\| = \|f\| = 1$. Therefore $\Phi \in D(A", \hat{1})$, and so

$$F(f) = \Phi(F) \in V(A", F) .$$

This gives

$$\{F(f) : f \in D(A, 1)\} \subset V(A", F) ,$$

and, since $V(A", F)$ is closed and convex, it completes the proof of the theorem.

The corollary is immediate.

Definition 4. Let A be a complex normed star algebra with a continuous involution $x \to x^*$. Mappings $f \to f^*$ and $F \to F^*$ are defined on A' and $A"$ respectively by

$$f^*(x) = (f(x^*))^* \quad (x \in A) ,$$

$$F^*(f) = (F(f^*))^* \quad (f \in A') .$$

It is clear that $F \to F^*$ is a mapping of $A"$ onto $A"$ that is antilinear and involutive, i. e.

$$(\alpha F + \beta F)^* = \alpha^* F^* + \beta^* G^*, \quad (F^*)^* = F .$$

However, it is not in general true that $(FG)^* = G^* F^*$. In fact, it is straightforward to verify that

$$(FG)^* = F^* \cdot G^* ,$$

where $F^* \cdot G^*$ denotes the Arens multiplication derived from the reversed multiplication on A. Thus A" with the Arens multiplication and with the natural involution $F \to F^*$ is a Banach star algebra if and only if the Arens multiplication is regular.

Theorem 5. Let A be a complex unital B*-algebra. Then the second dual space A" with the Arens multiplication and with the natural involution $F \to F^*$ is a complex unital B*-algebra.

Proof. Let $f \in D(A, 1)$, $F \in A$", and $F^* = F$. By Example 5. 3, f is a Hermitian functional on A, i. e.

$$f(a^*) = (f(a))^* \qquad (a \in A) .$$

Therefore $f^* = f$, and

$$F(f) = F^*(f) = (F(f^*))^* = (F(f))^* .$$

This proves that $F(f) \in \underset{\sim}{R}$, and so, by Corollary 3, F is a Hermitian element of A".

Each $F \in A$" can be written in the form $F = G + iH$ with $G = \frac{1}{2}(F + F^*)$, $H = \frac{1}{2i}(F - F^*)$, so that $G^* = G$, $H^* = H$. Thus every element of A" is a linear combination of Hermitian elements, and so, by the Vidav-Palmer theorem, A" is a B*-algebra with the Hermitian elements as its self-adjoint elements.

Remark. Let A be a complex B*-algebra with no unit element and let B be the B*-algebra obtained by adjoining a unit element to A. It is routine to verify that A" can be embedded isometrically and star isomorphically in B", so that A" is also a B*-algebra. Moreover, B" is the B*-algebra obtained by adjoining a unit element to A".

13. SPECTRAL STATES

Definition 1. Let A be a complex unital Banach algebra. We say that $f \in A'$ is a spectral state of A if

$$f(a) \in \mathrm{co}\, \mathrm{Sp}(A, a) \qquad (a \in A) .$$

We write $\Omega(A)$ for the set of all spectral states of A.

It is immediate from the definition that each spectral state f satisfies $f(1) = 1$ and $|f(a)| \leq \rho(a) \leq \|a\|$ $(a \in A)$, so that a spectral state is in particular a normalized state of A. The lemma below gives elementary characterizations of spectral states. Recall first from §2 that $\underline{N}$ denotes the set of all algebra-norms p on A that are equivalent to the given norm and satisfy $p(1) = 1$. Given $p \in \underline{N}$, let $D_p(A, 1)$ denote the set of normalized states of A with respect to the norm p. Since each $p \in \underline{N}$ determines the same topology on A, the sets $D_p(A, 1)$ are all subsets of A'.

Lemma 2. Let A be a complex unital Banach algebra and let $f \in A'$. Then the following statements are equivalent.

(i) $f \in \Omega(A)$.

(ii) $f(1) = 1$, $|f(a)| \leq \rho(a)$ $(a \in A)$.

(iii) $f \in D_p(A, 1)$ $(p \in \underline{N})$.

Proof. (i) → (ii). Clear.

(ii) → (iii) This follows from the fact that $\rho(a) \leq p(a)$ $(a \in A, p \in \underline{N})$. (iii) → (i) This follows from Theorem 2.10.

Remarks. (1) It is clear from Lemma 2(iii) that $\Omega(A)$ is a convex weak* compact subset of A'. However, $\Omega(A)$ can be empty as we shall show below.

(2) It is clear that $\Omega(A)$ contains all non-zero multiplicative linear functionals. Also, if A is commutative, then every probability measure on the carrier space of A gives rise to a spectral state on A. Let A be arbitrary and let f be a Jordan functional on A (i. e. $f(a^2) = f(a)^2$ $(a \in A)$) with $f(1) = 1$. It is elementary to show then that $|f(a)| \leq \rho(a)$ $(a \in A)$ so that $f \in \Omega(A)$. However, a result of Zelazko [77], which we discuss below, shows that such functionals are automatically multiplicative.

(3) If $\rho(a) = 0$, then $f(a) = 0$ for each $f \in \Omega(A)$. In particular, if R is the Jacobson radical of A, then $f(R) = \{0\}$ for each f in $\Omega(A)$. It is now easily verified that the canonical mapping $f \to \tilde{f}$ maps $\Omega(A)$ one-to-one onto $\Omega(A/R)$.

(4) Given a complex Banach algebra A without unit element we again say that a linear functional f on A is a <u>spectral state</u> of A if

$$f(a) \in \text{co } \text{Sp}(A, a) \qquad (a \in A).$$

Let $B = A + \underline{C}1$ be the unitization of A. Given a spectral state f of A, let $\hat{f}$ be the extension of f to B given by

$$\hat{f}((a, \alpha)) = f(a) + \alpha \qquad ((a, \alpha) \in B) .$$

Since

$$Sp(B, (a, \alpha)) = \alpha + Sp(B, (a, 0)) = \alpha + Sp(A, a)$$

it follows that $\hat{f}((a, \alpha)) \in \text{co } Sp(B, (a, \alpha))$ for each (a, α) in B. Thus $\hat{f}$ is a spectral state of B.

(5)   The concept of spectral state was suggested to us by R. T. Moore [45] under the name 'topological state' and the definition of Lemma 2(iii). We have preferred the name 'spectral state' since the set of them is determined by the algebraic structure rather than the topological structure of A. The following example illustrates this point.

Example 3.   Let A_1, A_2 denote $\underset{\sim}{C}^2$ with the multiplications given respectively by

$$(\alpha_1, \alpha_2)(\beta_1, \beta_2) = (\alpha_1\beta_1 + \alpha_1\beta_2 + \alpha_2\beta_1, \alpha_2\beta_2) ,$$

$$(\alpha_1, \alpha_2)(\beta_1, \beta_2) = (\alpha_1\beta_2 + \alpha_2\beta_1, \alpha_2\beta_2) ,$$

i. e. A_1 is the algebra obtained by adjoining a unit to $\underset{\sim}{C}$ with its usual multiplication, and A_2 is the algebra obtained by adjoining a unit to $\underset{\sim}{C}$ with the zero multiplication. Let A_1, A_2 be normed by

$$\|(\alpha_1, \alpha_2)\| = |\alpha_1| + |\alpha_2| .$$

Then A_1, A_2 are both complex unital Banach algebras with unit element $(0, 1)$. Let f be the functional defined on $\underset{\sim}{C}^2$ by

$$f((\alpha_1, \ \alpha_2)) = \alpha_1 + \alpha_2 \ .$$

Then f is multiplicative on A_1 and therefore a spectral state of A_1. On the other hand in A_2, $(1, 0)$ is nilpotent and $f((1, 0)) = 1$, so that f is not a spectral state of A_2.

Theorem 4. Let $f \in \Omega(A)$. Then

$$f(ab) = f(ba) \qquad (a, b \in A) \ .$$

Proof. Given $f \in \Omega(A)$, $a, b \in A$, let

$$F(\lambda) = f(\exp(-\lambda a)b \exp(\lambda a)) \qquad (\lambda \in \underset{\sim}{C}) \ .$$

Then F is an entire function and

$$\left| F(\lambda) \right| \leq \rho(\exp(-\lambda a)b \exp(\lambda a)) = \rho(b) \qquad (\lambda \in \underset{\sim}{C}) \ .$$

By Liouville's theorem, F is constant, and so

$$f(b) = f(\exp(-\lambda a)b \exp(\lambda a)) \qquad (\lambda \in \underset{\sim}{C}) \ . \tag{1}$$

Equate coefficients of λ in (1) to obtain $f(ab) = f(ba)$.

We can now give a simple example of a complex unital Banach algebra with no spectral states.

Example 5. Let $\mathcal{H}$ be an infinite-dimensional complex Hilbert space. Then $\Omega(B(\mathcal{H})) = \emptyset$.

Proof. Suppose that f is a spectral state of $B(\mathcal{H})$. By Halmos [30] Problem 186, Corollary 2, there exist $S, T, U, V \in B(\mathcal{H})$ such that

$$I = ST - TS + UV - VU .$$

It follows from Theorem 4 that $f(I) = 0$, which is impossible.

The above proof rests heavily on a rather difficult result for operators on Hilbert space. The alternative proof given below is simpler and of some interest. Given $a, b \in A$, we write $a \sim b$ if there exists an invertible element u in A such that $a = ubu^{-1}$.

Lemma 6. If A has elements a_1, a_2, a_3, a_4 such that $a_1 \sim a_2 \sim a_3 \sim a_4$ and $1 = a_1 + a_2$, $a_2 = a_3 + a_4$, then there does not exist any linear functional f on A with $f(1) = 1$ and such that $a \sim b$ implies $f(a) = f(b)$.

Proof. Suppose that f is such a functional. Then $a_1 \sim a_2$, $a_1 + a_2 = 1$ give $f(a_2) = \frac{1}{2}$. Also $a_2 \sim a_3 \sim a_4$ give $f(a_3) = f(a_4) = \frac{1}{2}$. This contradicts the fact that $a_2 = a_3 + a_4$.

Theorem 7. Let $\mathcal{H}$ be an infinite dimensional separable Hilbert space. Then there does not exist any linear functional f on $B(\mathcal{H})$ such that $f(I) = 1$ and $f(U A U^{-1}) = f(A)$ for all $A, U \in B(\mathcal{H})$, U invertible.

Proof. Let E_1, E_2 be closed infinite dimensional linear subspaces of $\mathscr{H}$ with infinite dimensional orthogonal complements, F_1, F_2, respectively. Then there exist linear homeomorphisms S, T of E_1 onto E_2 and F_1 onto F_2 respectively. For $i = 1, 2$, let P_i be the orthogonal projection onto E_i, Q_i onto F_i, so that $P_i + Q_i = I$, $P_i Q_i = Q_i P_i = 0$. Let

$$U = SP_1 + TQ_1, \quad V = S^{-1}P_2 + T^{-1}Q_2 .$$

Then $UV = VU = I$, and $UP_1 V = P_2$, so that $P_1 \sim P_2$.

It is now clear that $B(\mathscr{H})$ contains projections A_1, A_2, A_3, A_4 such that $A_1 \sim A_2 \sim A_3 \sim A_4$, $I = A_1 + A_2$, $A_2 = A_3 + A_4$. Lemma 6 now completes the proof.

It is now a corollary of Theorem 7 that $\Omega(B(\mathscr{H})) = \varnothing$. For, if f is a spectral state on $B(\mathscr{H})$ then $A = U\, BU^{-1}$ implies

$$f(A) = f(U\, BU^{-1}) = f(U^{-1}U\, B) = f(B) .$$

Theorem 8. (Gleason-Kahane-Zelazko.) Let A be a complex unital Banach algebra and let $f \in A'$. Then f is multiplicative if and only if

$$f(a) \in Sp(A, a) \quad (a \in A) .$$

Proof. It is well known that $f(a) \in Sp(A, a)$ $(a \in A)$ when f is multiplicative. Conversely, Gleason [26] and Kahane and Zelazko [36] proved independently that f is a Jordan functional whenever $f(a) \in Sp(A, a)$ $(a \in A)$. This gives the required result if A is commutative. Zelazko [77] extended the result to the general case. Alternatively we note that f is a spectral state and so $f(ab) = f(ba)$ $(a, b \in A)$ by Theorem 4. Since

$(a + b)^2 = a^2 + ab + ba + b^2$ and $f(x^2) = f(x)^2$ $(x \in A)$, it follows that f is multiplicative.

For the next two results, A need not be complete nor unital. Given $u \in X$, $\phi \in X'$, $u \otimes \phi$ denotes the operator defined by

$$(u \otimes \phi)(x) = \phi(x)u \quad (x \in X) .$$

Lemma 9. Let X be a complex normed space, A a subalgebra of $B(X)$ containing all finite rank operators, and f a linear functional on A such that $|f(a)| \leq \rho(a)$ $(a \in A)$. Then there exists a constant λ with $|\lambda| \leq 1$ such that

$$f(u \otimes \phi) = \lambda\phi(u) \quad (u \in X, \ \phi \in X') .$$

Proof. Let $u \in X$, $\phi \in X'$, $a = u \otimes \phi$. Then $a^2 = \phi(u)a$ and so $\rho(a) = |\phi(u)|$. Therefore

$$|f(u \otimes \phi)| \leq |\phi(u)| \quad (u \in X, \ \phi \in X') . \tag{2}$$

Let F_ϕ be defined on X by

$$F_\phi(x) = f(x \otimes \phi) \quad (x \in X) .$$

Then $F_\phi \in X'$ and it follows from (2) that the null space of F_ϕ contains the null space of ϕ. Hence there is $\lambda_\phi \in \underset{\sim}{C}$ with $|\lambda_\phi| \leq 1$ such that $F_\phi = \lambda_\phi\phi$.

Let ϕ, ψ be linearly independent elements of X'. Then

$$\lambda_{\phi+\psi}(\phi + \psi) = F_{\phi+\psi} = F_\phi + F_\psi = \lambda_\phi\phi + \lambda_\psi\psi. \qquad (3)$$

Also, given $\alpha \in \underset{\sim}{C} \setminus \{0\}$, $\phi \in X' \setminus \{0\}$, we have

$$\lambda_{\alpha\phi}(\alpha\phi) = F_{\alpha\phi} = \alpha F_\phi = \alpha\lambda_\phi\phi. \qquad (4)$$

It follows from (3) and (4) that λ_ϕ is constant for $\phi \in X'$.

Theorem 10. Let X be an infinite-dimensional normed space, let A be a subalgebra of $B(X)$ containing all finite rank operators, and let f be a linear functional on A such that $|f(a)| \leq \rho(a)$ $(a \in A)$. Then $f(a) = 0$ for all finite rank operators a.

Proof. Let n be a positive integer and let $u_1, \ldots, u_n$ be linearly independent elements of X. Choose $\phi_1, \ldots, \phi_n \in X'$ such that

$$\phi_j(u_i) = \begin{cases} 1 & \text{if } i = j, \\ 0 & \text{if } i \neq j. \end{cases}$$

Let $a = u_1 \otimes \phi_1 + \ldots + u_n \otimes \phi_n$. Then $a^2 = a$ and $\rho(a) = 1$. Let λ be as in Lemma 9. Then

$$1 \geq |f(a)| = |\lambda(1 + \ldots + 1)| = n|\lambda|.$$

Since n is arbitrary, $\lambda = 0$, and the result follows from Lemma 9.

We give next a complete description of the spectral states on a finite dimensional complex unital Banach algebra. Let A be

any finite dimensional complex linear algebra with unit element 1.
By the Wedderburn theorem, we have

$$A = A_1 \oplus \ldots \oplus A_m \oplus R \tag{5}$$

where R is the radical of A and each A_j is a subalgebra of A
that is isomorphic with a full matrix algebra over $\underset{\sim}{C}$. We identify
each A_j with the corresponding full matrix algebra, and we write
T_j for the normalized trace on A_j (the normalized trace of an
$n \times n$ matrix (a_{pq}) being $\frac{1}{n}(a_{11} + \ldots + a_{nn})$). We also write
T_j for the natural extension of T_j to A given by

$$T_j(a) = T_j(a_j)$$

where $a = a_1 + \ldots + a_m + r$ is the decomposition of a given by
(5).

> **Theorem 11.** Let A be a finite dimensional complex
> unital Banach algebra with Wedderburn decomposition
> given by (5). Then $\Omega(A)$ is the convex hull of the normalized
> traces T_j $(j = 1, \ldots, m)$.

Proof. Recall that the sum of the eigenvalues of a matrix
is given by the trace of the matrix, and hence the normalized trace
of a matrix is dominated by the spectral radius of the matrix. For
$j = 1, \ldots, m$, let e_j be the central projection on A_j. Then
$T_j(1) = T_j(e_j) = 1$, and

$$|T_j(a)| = |T_j(ae_j)| \leq \rho(ae_j) \leq \rho(a)\rho(e_j) = \rho(a) \quad (a \in A) .$$

Lemma 2 now gives $T_j \in \Omega(A)$.

Given any $f \in \Omega(A)$ we have $f(R) = \{0\}$, since each element of R is nilpotent. Let A_j consist of all complex $p \times p$ matrices. Then the restriction of f is a linear functional on A_j and

$$|f(b)| \le \rho(b) \quad (b \in A_j) .$$

Let $X = \underset{\sim}{C}^p$ with natural basis $u_1, \ldots, u_p$ and dual basis $\phi_1, \ldots, \phi_p$. By Lemma 9, there is $\lambda_j \in \underset{\sim}{C}$ with

$$f(u_r \otimes \phi_s) = \lambda_j \phi_s(u_r) .$$

In particular,

$$f(u_r \otimes \phi_s) = 0 \quad (r \ne s)$$

$$f(u_r \otimes \phi_r) = \lambda_j \quad (1 \le r \le p) .$$

Given $b \in A_j$, we have $b = \Sigma_{r,s} b_{rs} u_r \otimes \phi_s$ and so

$$f(b) = \lambda_j \Sigma_r b_{rr} = p\lambda_j T_j(b) = f(e_j)T_j(b) .$$

It follows that

$$f(a) = \sum_{j=1}^{m} f(e_j)T_j(a) \quad (a \in A) .$$

Since $f(e_j) \in co\ Sp(A, e_j)$, we have $0 \le f(e_j) \le 1$. Also

$$1 = f(1) = f(e_1 + \ldots + e_m) = f(e_1) + \ldots + f(e_m) .$$

Therefore $f \in co\{T_j : j = 1, \ldots, m\}$ and the proof is complete.

Given an arbitrary complex unital Banach algebra A, it is natural to seek irreducible representations of A associated with spectral states of A. We recall first some notation from the theory of dual representations (see Bonsall and Duncan [13]). Given $f \in A'$, we write

$$L_f = \{x : f(Ax) = \{0\}\}$$

$$K_f = \{x : f(xA) = \{0\}\}$$

$$P_f = \{x : f(AxA) = \{0\}\}$$

$$X_f = A - L_f, \quad Y_f = A - K_f$$

$$< x', y' >_f = f(yx) \quad (x \in x' \in X_f, \; y \in y' \in Y_f)$$

$$T_a^f x' = (ax)' \quad (x \in x' \in X_f, \; a \in A)$$

$$S_a^f y' = (ya)' \quad (y \in y' \in Y_f, \; a \in A) .$$

We recall that $(X_f, Y_f, < , >_f)$ are Banach spaces in normed duality and $a \to T_a^f$ is a dual representation of A on $(X_f, Y_f, < , >_f)$ with $(T_a^f)^* = S_a^f$.

Lemma 12. Let $f \in \Omega(A)$. Then $L_f = K_f = P_f$.

Proof. It is immediate from Theorem 4 that $L_f = K_f$. Since A is unital, we have

$$L_f \supset P_f = \{x : Ax \subset K_f\} = \{x : Ax \subset L_f\} \supset L_f .$$

The proof is complete.

Lemma 13. Let X be a complex Banach space and let A be a non-zero strictly irreducible subalgebra of B(X) with no left ideals except {0} and A. Then dim(X) = 1.

Proof. Since X is complex, A is strictly dense by Rickart [57] Theorem (2.4.6). Given $x \neq 0$, {T ∈ A : Tx = 0} is a proper left ideal of A and hence is {0}. Thus $x \neq 0$, Tx = 0 implies T = 0. Since A is strictly dense, it follows that dim(X) = 1.

Theorem 14. Given f ∈ Ω(A), the following statements are equivalent.

(i) $a \to T_a^f$ is dually strictly irreducible.

(ii) $a \to T_a^f$ or $a \to S_a^f$ is strictly irreducible.

(iii)   f is multiplicative.

Proof. (i) → (ii). Trivial.

(ii) → (iii). Let $a \to T_a^f$ be strictly irreducible. Then L_f is a maximal left ideal. It follows from Lemma 12 that A/P_f is a primitive Banach algebra with no non-zero proper left ideals. It now follows from Lemma 13 that A/P_f is one dimensional. Since f(1) = 1, we conclude that f is multiplicative. A similar argument applies if $a \to S_a^f$ is strictly irreducible.

(iii) → (i). Elementary.

Remark. We have f ∈ Ω(A) if and only if f(a) ∈ co Sp(A, a) for each a ∈ A. We may say that f is 'extreme' if f(a) ∈ Sp(A, a) for each a ∈ A. Theorem 8 shows that this condition characterizes the multiplicative linear functionals. Thus the 'extreme' spectral states correspond to the irreducible spectral states. It should be noted that the 'extreme' spectral states are not the same as the extreme points of the set of all spectral states. For example, in the notation of Theorem 11, the extreme points of $\Omega(A)$ are the normalized traces T_j, and a normalized trace is multiplicative if and only if the matrix algebra is one dimensional.

14. REMARKS AND PROBLEMS

Remarks. (1) We have not discussed any of the relations between the theory of numerical range and the theory of spectral operators on a Banach space. For example, Berkson [6] has shown that, if T is a bounded operator on a reflexive Banach space, then T is a spectral operator of scalar type if and only if $T = R + iJ$ where $RJ = JR$ and, for some equivalent renorming on X, $R^m J^n$ $(m, n = 0, 1, 2, \ldots)$ are Hermitian. Lumer [42] has shown that a Hermitian operator may fail to be spectral in general. In the same paper, Lumer shows that if the uniformly closed algebra generated by a Boolean algebra of projections on a reflexive Banach space consists of spectral operators of scalar type, then the Boolean algebra is uniformly bounded. This shows that one of Dunford's original conditions on admissible spectral measures is necessary for an adequate theory of spectral operators. For details and further topics, the reader should consult the above papers together with Berkson [7], Palmer [50], and Panchapagesan [52].

(2) Lumer [41] has used numerical range techniques to characterize the isometries of certain reflexive Orlicz spaces.

(3)   Lumer (private communication) has recently shown that there exist constants c_1, c_2 such that, for real unital Banach algebras A,

$$\|a\| \le c_1 \, v(a) + c_2 (v(a^2))^{\frac{1}{2}} \quad (a \in A) .$$

For real C*-algebras we may take $c_1 = 2$, $c_2 = 1$. M. J. Crabb (Ph. D. thesis) has shown that if A is a real unital Banach algebra, then

$$\|a\| \le \max \{48v(a), \ 24v(a^2)^{\frac{1}{2}}\} \quad (a \in A) . \ ^{†}$$

(4)   Sinclair [63] has recently shown that, if A is a unital B*-algebra, then $T \in B(A)$ is a Hermitian operator if and only if T is the sum of a left multiplication by a self-adjoint element of A and a star derivation of A.

(5)   It is shown in Cudia [18] that the mapping $x \rightarrow D(X, x)$ is upper semi-continuous when S(X) has the norm topology and X' the weak* topology. It is of interest to consider the situation when X' has its norm topology. For this case, Bonsall, Cain, and Schneider [12] show that $x \rightarrow D(X, x)$ is upper semi-continuous for $X = c_0$, but not for any subspace of ℓ_∞ that contains c, nor for $X = \ell_1$.

(6)   It is shown in Theorem 11. 4 that the set

$$\Pi = \{(x, f) : \|x\| = 1 = f(x) = \|f\| \}$$

is connected in the product topology when X has the norm topology and

---

† $\|a\| \le 4 \max \{v(a), v(a^2)^{\frac{1}{2}}\}$ (C. M. McGregor).

X' has the weak* topology and X is not $\underset{\sim}{R}$. Again it is of interest to consider the situation when X' has the norm topology. For this case it is not difficult to show that Π is connected if X has smooth unit ball (and $X \neq \underset{\sim}{R}$) or if $X = \ell_1$. M. J. Crabb (private communication) has shown that Π is connected for certain subspaces X of C(E) where E is a compact Hausdorff space. To be precise, Π is connected if, given $f, g \in X$ with $\|f\| = \|g\| = 1$, there exist $s, t \in E$ and $h \in X$ such that $|f(s)| = |h(s)| = |g(t)| = |h(t)| = \|h\| = 1$.

(7)   Let X be a normed space and let A be a subalgebra of B(X). Then each element (x, f) of Π (as in (6)) induces an evaluation functional on A given by

$$F(T) = f(Tx) \quad (T \in A) .$$

Let $\hat{\Pi}$ denote the set of all such evaluation functionals. It is proved in Duncan [20] that if X is complete and A contains all operators of finite rank then $\hat{\Pi}$ is norm closed in A' though it need not be weak* closed.

(8)   Palmer [51] has recently proved, by numerical range techniques, that a real Banach star algebra is C* if and only if it is B* with Hermitian involution (i. e. each self-adjoint element has real spectrum).

(9)   M. J. Crabb (private communication) has very recently constructed an example of a four dimensional complex commutative unital Banach algebra with Hermitian elements u, v such that $\rho(u + iv) = \frac{1}{\sqrt{2}} \|u + iv\|$. This shows that Sinclair's result on the

norm of Hermitian elements does not extend to normal elements.

(10) M. J. Crabb (private communication) has shown that the inequality in Theorem 4. 8 is best possible in the following strong sense. There is a complex unital Banach algebra A and non-zero a $\in$ A such that

$$\|a^n\| = n! \, (e/n)^n v(a)^n \quad (n = 1, 2, \dots) .$$

It is an immediate corollary of the above that the power inequality fails for the element a. The above algebra has numerical index e^{-1}. C. M. McGregor (private communication) has shown that, given $1 > t \geq e^{-1}$, there is a two dimensional complex normed space X such that $n(X) = t$ and the power inequality fails for some T $\in$ B(X).

(11) Given a Banach space X, we have $n(X) = 1$ if and only if the numerical radius is an algebra norm on B(X). The necessity is trivial and the sufficiency follows from Theorem 8 of F. F. Bonsall, 'A minimal property of the norm in some Banach algebras', J. London Math. Soc. 29 (1954), 156-164.

(12) Let A be a real or complex unital Banach algebra. C. M. McGregor (private communication) has shown that if $u \in A$, $u^2 = u$, and $v(u) = 1$, then $\|u\| = 1$.

(13) Given a finite dimensional complex normed space X, Zenger [78] has shown that co Sp(T) $\subset$ V(T) for each T $\in$ B(X). M. J. Crabb (Ph. D. thesis) has shown that we have co Sp(T) $\subset$ $V(T)^-$ for each T $\in$ B(X) where X is any complex Banach space.

(14)   A. M. Sinclair (preprint) has shown that if $0 \in \partial \overline{\mathrm{co}}\ V(T)$, then the kernel of T is orthogonal to the range of T. (A linear subspace Y of X is said to be <u>orthogonal</u> to a linear subspace Z of X if

$$\|y\| \leq \|y + z\| \quad (y \in Y, \ z \in Z).)$$

(15)   M. J. Crabb (Ph. D. thesis) has given an example in which $V(T) \neq V(T^*)$. B. Bolobbas (preprint) has extended the Bishop-Phelps theorem to show that $V(T)^- = V(T^*)^-$ $(T \in B(X))$ for any Banach space X.

(16)   M. J. Crabb (Ph. D. thesis and private communication) has established the following results on Hermitian elements.   Let $h \in H(A)$.   Then $V(h^2) \subset \{z : \mathrm{Re}\ z \geq 0\}$.   Now let $\|h\| = 1$.   Then, for all real t with $|t - \frac{\pi}{2}| \leq \frac{\pi}{9}$,

$$\|\cos t + ih^3 \sin t\| \leq 1 .$$

Let $f \in D(A, h)$.   Then

$$f(h^{2n-1}) = 1 \quad (n = 1, 2, \ldots)$$

and there is some real k such that

$$f(h^{2n}) = k \quad (n = 0, 1, 2, \ldots) .$$

If, further, $f(1) = 1$, then $f(h^n) = 1$ $(n = 1, 2, 3, \ldots)$ and f is multiplicative on the closed subalgebra of A generated by 1 and h.

(17)   Further material of interest may be found in references [25], [44], [65], [75], [76].

Problems.   In (1) - (6),   A denotes a complex unital Banach algebra.

(1)   What special numerical range features occur if A has minimal norm i. e. if $|.| = \|.\|$ whenever $|.|$ is an algebra-norm on A such that

$$|a| \leq \|a\| (a \in A) ?$$

(2)   Let A be such that a $\in$ A is Hermitian whenever a has real spectrum. Several people have shown that A has no quasinilpotents (and hence is semi-simple) and that J(A) is in the centre of A. What further can be said about the structure of A?

(3)   Can one give a numerical range proof of the Shirali-Ford theorem [61]?

(4)   Develop a satisfactory theory for Hermitian elements h whose powers are not all Hermitian. Is the spectral radius an algebra-norm on the smallest closed subalgebra of A containing h?

(5)   Evaluate α where

$$\alpha = \inf \{\rho(u): u \in S(A),   u \text{ normal, all } A\} .$$

It is easy to show that $\alpha \geq \frac{1}{2}$ and M. J. Crabb (private communication has shown that $\alpha < 1/\sqrt{2}.$[†] Evaluate β where

---

[†] $\alpha = \frac{1}{2}$.

$\beta = \inf\{\rho(u) : u \in S(A),\ u \text{ normal, Re } u,\ \text{Im } u \in K(A),\ \text{all } A\}.$

C. M. McGregor (private communication) has shown that $\beta \geq 1/\sqrt{2}$.

(6)   Let A be commutative. Characterize those A such that $\text{ext}\Omega(A)$ is the set of multiplicative linear functionals on A.

In (7) - (14), X denotes a normed space and $T \in B(X)$.

(7)   Is $V(T)$ always simply connected? Is $V(T)^-$ always simply connected? The two questions coincide for finite dimensional spaces and C. M. McGregor (private communication) has shown that the answer is 'yes' for absolute norms on $\underset{\sim}{C}^2$. Is $V(T)$ always arcwise connected? The answer is known to be 'yes' for various classes of normed spaces including smooth spaces.

(8)   For what operators T and spaces X is $V(T)$ a closed set?

(9)   Let K be a compact simply connected subset of $\underset{\sim}{C}$. Is there some X and $T \in B(X)$ such that $V(T) = K$ or $V(T)^- = K$?

(10)   Characterize those normed spaces X such that $V(T)$ is convex for every $T \in B(X)$. (See Zenger [78].)

(11)   Characterize those normed spaces X such that the power inequality holds in $B(X)$.

(12)   For each positive integer n determine

$$\sup\{v(T^n) : T \in B(X),\ v(T) = 1,\ \text{all } X\}.^\dagger$$

$\dagger$ $\sup\{f^{(n)}(0): f \text{ entire},\ f(0)=1,\ |f(z)| \leq e^{|z|}\ (z \in \underset{\sim}{C})\}$ (M. J. Crabb, J. G. Clunie).

(13) Is every point λ of the topological boundary $\partial V(T)$ accessible from the complement of $V(T)$, in the sense that the complement of $V(T)$ contains an open sector of a disc with its vertex at λ ?

(14) Classify A*-algebras (without unit element in general) according to the size of the set of Hermitian elements H with respect to the given Banach algebra-norm (or some equivalent norm). Consider in particular the cases when $H = \{0\}$ and $A = H + iH$.

(15) Characterize those complex Banach spaces which are B*-algebras for some multiplication and some involution (c.f. the remarks at the end of §6).

(16) Johnson [34] has shown that every semi-simple Banach algebra has unique complete norm topology. On the other hand, non-semi-simple Banach algebras can have non-equivalent Banach-algebra-norms. Compare the numerical ranges with respect to non-equivalent norms.

BIBLIOGRAPHY

1. R. Arens, 'Operations induced in function classes', Monat. für Math. 55 (1951) 1-19.

2. R. Arens, 'The adjoint of a bilinear operation', Proc. Amer. Math. Soc. 2 (1951) 839-848.

3. F. L. Bauer, 'On the field of values subordinate to a norm', Numer. Math. 4 (1962) 103-111.

4. C. A. Berger, 'On the numerical range of an operator', Bull. Amer. Math. Soc. (to appear).

5. C. A. Berger and J. G. Stampfli, 'Mapping theorems for the numerical range', Amer. J. Math. 89 (1967) 1047-1055.

6. E. Berkson, 'A characterization of scalar type operators on reflexive Banach spaces', Pac. J. Math. 13 (1963) 365-373.

7. E. Berkson, 'Some types of Banach spaces, Hermitian operators, and Bade functionals', Trans. Amer. Math. Soc. 116 (1965) 376-385.

8. E. Berkson, 'Some characterizations of C*-algebras', Ill. J. Math. 10 (1966) 1-8.

9. E. Bishop and R. R. Phelps, 'A proof that every Banach space is subreflexive' Bull. Amer. Math. Soc. 67 (1961) 97-98.

10. H. F. Bohnenblust and S. Karlin, 'Geometrical properties of the unit sphere of Banach algebras', Ann. of Math. 62 (1955) 217-229.

11. F. F. Bonsall, 'The numerical range of an element of a normed algebra', Glasgow Math. J. 10 (1969) 68-72.

12. F. F. Bonsall, B. E. Cain and H. Schneider, 'The numerical range of a continuous mapping of a normed space', Aequationes Math. 2 (1968) 86-93.

13. F. F. Bonsall and J. Duncan, 'Dual representations of Banach algebras', Acta Math. 117 (1967) 79-102.

14. F. F. Bonsall and J. Duncan, 'Dually irreducible representations of Banach algebras', Q. J. Math. (Oxford) 19 (1968) 97-111.

15. P. Civin and B. Yood, 'The second conjugate space of a Banach algebra as an algebra', Pac. J. Math. 11 (1961) 847-870.

16. J. A. Clarkson, 'Uniformly convex spaces', Trans. Amer. Math. Soc. 40 (1936) 396-414.

17. M. J. Crabb, 'Numerical range estimates for the norms of iterated operators' (to appear in Glasgow Math. J.).

18. D. F. Cudia, 'The geometry of Banach spaces. Smoothness', Trans. Amer. Math. Soc. 110 (1964) 284-314.

19. M. M. Day, 'Normed linear spaces', Springer-Verlag (1958).

20. J. Duncan, 'The evaluation functionals associated with an algebra of bounded operators', Glasgow Math. J. 10 (1969) 73-76.

21. J. Duncan, C. M. McGregor, J. D. Pryce and A. J. White, 'The numerical index of a normed space', (to appear in J. London Math. Soc.).

22. J. M. G. Fell, 'The dual spaces of Banach algebras', Trans. Amer. Math. Soc. 114 (1965) 227-250.

23. F. R. Gantmacher, 'Matrix theory', volume II, Chelsea (1959).

24. I. M. Gelfand and M. A. Naimark, 'On the embedding of normed rings into the ring of operators in Hilbert space', Mat. Sbornik 12 (1943) 197-213.

25. J. R. Giles, 'Classes of semi-inner-product spaces', Trans. Amer. Math. Soc. 129 (1967) 436-446.

26. A. M. Gleason, 'A characterization of maximal ideals', J. Analyse Math. 19 (1967) 171-172.

27. B. W. Glickfeld, 'A metric characterization of $C(X)$ and its generalization to C*-algebras', Ill. J. Math. 10 (1966) 547-566.

28. B. W. Glickfeld, 'On an inequality of Banach algebra geometry and semi-inner-product space theory', (to appear in Ill. J. Math.).

29. J. G. Glimm and R. V. Kadison, 'Unitary operators in C*-algebras', Pac. J. Math. 10 (1960) 547-556.

30. P. R. Halmos, 'Hilbert space problem book', Van Nostrand (1967).

31. F. Hausdorff, 'Der Wertvorrat einer Bilinearform', Math. Zeit. 3 (1919) 314-316.

32. R. A. Hirschfeld and W. Zelazko, 'On spectral norm Banach algebras', Bull. Acad. Pol. Sci. 16 (1968) 195-199.

33. R. B. Holmes, 'A formula for the spectral radius of an operator', Amer. Math. Monthly 75 (1968) 163-166.

34. B. E. Johnson, 'The uniqueness of the (complete) norm topology', Bull. Amer. Math. Soc. 73 (1967) 537-539.

35. R. V. Kadison, 'Isometries of operator algebras', Ann. of Math. 54 (1951) 325-338.

36. J.-P. Kahane and W. Zelazko, 'A characterization of maximal ideals in commutative Banach algebras', Studia Math. 29 (1968) 339-340.

37.　I. Kaplansky, 'Normed algebras', Duke Math. J. 16 (1949) 399-418.

38.　T. Kato, 'Some mapping theorems for the numerical range', Proc. Japan Acad. 41 (1965) 652-655.

39.　C. Le Page, 'Sur quelques conditions entrainant la commutativité dans les algèbres de Banach', Comptes Rendues (Paris), 265 (1967) 235-237.

40.　G. Lumer, 'Semi-inner-product spaces', Trans. Amer. Math. Soc. 100 (1961) 29-43.

41.　G. Lumer, 'Isometries of reflexive Orlicz spaces', Ann. Inst. Fourier (Grenoble) 13 (1963) 99-109.

42.　G. Lumer, 'Spectral operators, Hermitian operators and bounded groups', Acta Sci. Math. (Szeged) 25 (1964) 75-85.

43.　G. Lumer and R. S. Phillips, 'Dissipative operators in a Banach space', Pac. J. Math. 11 (1961) 679-698.

44.　R. T. Moore, 'Operator theory on locally convex spaces I: Banach algebras, states and numerical ranges', (in preparation).

45.　R. T. Moore, 'Approximation of spectra by numerical ranges (to appear).

46.　B. Sz. Nagy and C. Foias, 'On certain classes of power bounded operators in Hilbert space', Acta Szeged 27 (1966) 17-25.

47.　N. Nirschl and H. Schneider, 'The Bauer fields of values of a matrix', Numer. Math. 6 (1964) 355-365.

48.　T. Ono, 'Note on a B*-algebra', J. Math. Soc. Japan 11 (1959) 146-158.

49.　T. W. Palmer, 'Characterizations of C*-algebras', Bull. Amer. Math. Soc. 74 (1968) 538-540.

50.   T. W. Palmer, 'Unbounded normal operators on Banach spaces', Trans. Amer. Math. Soc. 133 (1968) 385-414.

51.   T. W. Palmer, Notices Amer. Math. Soc. 16 (1969) 221.

52.   T. V. Panchapagesan, 'Unitary operators in Banach spaces', Pac. J. Math. 22 (1967) 465-475.

53.   A. L. T. Paterson, 'Isometries between B*-algebras', (to appear in Proc. Amer. Math. Soc.).

54.   C. Pearcy, 'An elementary proof of the power inequality for the numerical radius', Mich. Math. J. 13 (1966) 289-291.

55.   G. Polya and G. Szegö, 'Aufgaben und Lehrsätze aus der Analysis' (Erster Band), Springer-Verlag (1954).

56.   C. R. Putnam, 'Commutation properties of Hilbert space operators and related topics', Ergebnisse der Mathematik und ihrer Grenzgebiete, Band 36, Springer-Verlag (1967).

57.   C. E. Rickart, 'General theory of Banach algebras', Van Nostrand (1960).

58.   B. Russo and H. A. Dye, 'A note on unitary operators in C*-algebras', Duke Math. J. 33 (1966) 413-416.

59.   I. E. Segal, 'Irreducible representations of operator algebras', Bull. Amer. Math. Soc. 53 (1947) 73-88.

60.   S. Sherman, 'The second adjoint of a C*-algebra', Proc. International Congr. Math. , Cambridge 1 (1950) 470.

61.   S. Shirali and J. W. M. Ford, 'Symmetry in complex involutory Banach algebras II' (to appear in Duke Math. J.).

62.   A. M. Sinclair, 'The norm of a Hermitian element in a Banach algebra' (to appear).

63.   A. M. Sinclair, 'Jordan homomorphisms and derivations on semi-simple Banach algebras' (to appear).

64.   I. N. Spatz, 'Geometry of Banach algebras', Notices Amer. Math. Soc. 14 (1967) # 643-15.

65. J. G. Stampfli, 'Minimal range theorems for operators with thin spectra', Pac. J. Math. 23 (1967) 601-612.

66. J. G. Stampfli, 'An extreme point theorem for inverses in a Banach algebra with identity', Proc. Camb. Phil. Soc. 63 (1967) 993-994.

67. J. G. Stampfli and J. P. Williams, 'Growth conditions and the numerical range in a Banach algebra', Tôhoku Math. J. 20 (1968) 417-424.

68. M. H. Stone, 'Linear transformations in Hilbert space', Amer. Math. Soc. Coll. Publ. XV (1932).

69. Z. Takeda, 'Conjugate spaces of operator algebras', Proc. Japan Acad. 30 (1954) 90-95.

70. O. Toeplitz, 'Das algebraische Analogon zu einem Satze von Fejer', Math. Zeit. 2 (1918) 187-197.

71. M. Tomita, 'The second dual of a C*-algebra', Mem. Fac. Sci. Kyushu Univ. Ser. A 21 (1967) 185-193.

72. I. Vidav, 'Eine metrische Kennzeichnung der selbstadjungierten Operatoren', Math. Zeit. 66 (1956) 121-128.

73. B. J. Vowden, 'On the Gelfand-Neumark theorem', J. London Math. Soc. 42 (1967) 725-731.

74. J. P. Williams, 'Spectra of products and numerical ranges', J. Math. Anal. and Appl. 17 (1967) 214-220.

75. J. P. Williams, 'Schwarz norms for operators', Pac. J. Math. 24 (1968) 181-188.

76. J. P. Williams and T. Crimmins, 'On the numerical radius of a linear operator', Amer. Math. Monthly 74 (1967) 832-833.

77. W. Zelazko, 'A characterization of multiplicative linear functionals in complex Banach algebras', Studia Math. 30 (1968) 83-85.

78.   Chr. Zenger, 'On convexity properties of the Bauer field of values of a matrix', Num. Math. 12 (1968) 96-105.

INDEX

A' 14

$A_{\underset{\sim}{C}}$ 19

$A(\Delta)$ 88

A*-algebra 73

adjoint (of an operator) 85, 127

algebra,

    unital normed 14

    convolution measure 79

approximate unit 76

approximate point spectrum 6, 92

Arens multiplication 9, 106, 107

ascent 6, 94, 96

auxiliary norm 73

B*-algebra 2, 5, 9, 41, 43, 44, 67-72, 78, 110, 124

B_{o}^{*}-algebra 67

Bauer 5, 100

Berger 3, 41

Berkson 57, 123

$B(\mathscr{H})$ 115

Bishop 84, 89, 90

Bohnenblust 5, 7, 30, 34, 37

Bollobas 127

Bonsall 6, 10, 26, 36, 101, 121, 124, 126

boundary 92, 94, 130

bounded multiplicative semi-group 21, 30, 90

B(X) 78, 81

$\underset{\sim}{C}$ 14

Cain 6, 101, 124

C*-algebra 9, 56, 65, 68

C(E) 70, 87

$C_{o}(G)$ 79

Civin 106, 107

Clarkson 92, 93

co 20

$\overline{co}$ 83

commutator 105

completion 17, 73

complexification 19, 24, 55

connectedness

    of Π 101

    of V(T) 6, 102

convex hull 20, 53, 54, 91, 126

    closed 83

convex, strictly   93
      uniformly   92
convexity of V(T)   6, 98, 100
convolution measure algebra   79
Crabb   7, 10, 26, 41, 57, 94, 97, 124-128
Cudia   124

∂ 91
D(A, x)   14
Day   92, 93
disc algebra   88
dissipative   30, 95
$D_p(A, 1)$ 111
dual representation   121
dual space   14, 81
Duncan   7, 10, 36, 87, 121, 125
Dunford   123
Dye   64

eigenvalue   2, 6, 89, 93, 94, 96
equivalent norm   20
exp   26
exponential function   26
extreme point   36, 40, 88, 123

$\underset{\sim}{F}$ 14
Φ_C 63
Fell   10
finite rank operators   117
Foias   41

Ford   61, 128
full matrix algebra   119

G(A)   26
$G_1(A)$ 27
Gantmacher   96
Gelfand   56, 68
Gleason   116
Glickfeld   34, 57, 87
Glimm   9, 68

H(A)   46
H*-algebra   77
Halmos   1, 3, 41, 44, 87, 115
Harris   64
Hausdorff   2
Hermitian,
      element   8, 9, 46, 58, 78, 94, 95, 108, 123, 124, 127, 128
      equivalent   47, 57
      functional   65, 69
$\mathcal{H}_f$ 65
Hilbert space   1
Hilbert-Schmidt operators   77
Hirschfeld   32
Holmes   91

index,
    numerical   7, 43, 87,
    126
    of eigenvalue   6, 94, 96
invertible   18
involution   50, 59, 109
irreducible algebra of operators
    122
isometries between B*-algebras
    72, 73

J(A)   49
Johnson   130
joint,
    numerical range   23
    spectrum   24
Jordan product   60

K(A)   48
Kadison   9, 68, 72
Kahane   116
Kaplansky   9, 61, 72
Karlin   5, 7, 30, 34, 37
Kato   41

Le Page   32
$L_1(G)$ 77
$\ell_1(G)$ 75
local uniform convexity   36
L_p 88

ℓ_p 78
L-space   87
Lumer   3-6, 8, 10, 17, 30, 34,
    37, 43, 47, 50, 55, 57,
    69, 72, 85, 87, 91, 92,
    123

maximal ideal space   55, 112
max Re Sp(a)   32, 51
max Re V(a)   17, 28, 51
McGregor   7, 87, 88, 126, 129
meromorphic operator   97
M(G)   79
Moore   10, 113
M-space   87

$\underline{N}$ 20
n(A)   43
Nagy   41
Naimark   56, 68
nilpotent   120
Nirschl   6, 7, 94, 100
norm × weak* topology   100, 125
normal element   54, 126, 128
normal cone   48
normalized state   14
normalized trace   119
numerical index   7, 43, 87,
    126
numerical radius   15, 34

numerical range   8, 15, 51,
    54, 108
    joint   23
    spatial   5, 81
n(X)   87

$\Omega(A)$ 111
Ono   68
Operators,
    Hilbert-Schmidt   77
    meromorphic   97
    spectral   123
Orlicz space   123

Palmer   9, 57, 60, 64, 65,
    77, 123, 125
Panchapagesan   123
Paterson   73
Pearcy   3
Phelps   84, 89, 90
Phillips   30
Pickford   77
positive   48
power inequality   41, 126
pre-B*-algebra   73-76
principal component of G(A)   27,
    61
Pryce   7, 87
Putman   3
$\Pi(X)$, Π 81

quasinilpotent   30

$\underset{\sim}{R}$ 14
$\rho(a)$ 19
regular Arens multiplication
    107
Rickart   19, 20, 27, 61, 65, 72,
    77, 79, 122
rotund,   93
    uniformly   92
Russo   64

S(A)   14
Schneider   6, 7, 94, 100, 101,
    124
second dual (of a normed algebra)
    106
Segal   72
semi-group, bounded multiplica-
    tive   21, 30, 90
semi-inner product (s. i. p.)   3,
    85, 103
Sherman   9, 106
Shirali   61, 128
simply connected   104, 129
Sinclair   10, 54, 61, 77, 124, 127
$Sp(A, a)$ 18
$Sp(A; a_1, \ldots, a_n)$ 24
spatial numerical range   5, 81
Spatz   56
spectral operators   123

141

spectral radius   19, 32, 61

spectral state   10, 111

spectrum   18, 19, 23, 51, 54,
      88, 91, 126
      joint   24

Stampfli   30, 39, 41

star algebra   59

state (normalized)   14

strictly convex   93

strictly irreducible algebra   122

stochastic   96

Stone   2

support functional   14

S(X)   81

symmetric   61

Takeda   106

Toeplitz   1, 2

Tomita   106

topological boundary   92, 94, 130

Torrance   77

total   36

trace   119

uniformly convex   92

uniformly rotund   92

unital normed algebra   14

upper semi-continuous   20, 124

$V(A, a)$ 15

$V(A; a_1, \ldots, a_n)$ 23

$V(A, a, x)$ 15

$V(a)$ 15

$v(a)$ 15

vertex   36

Vidav   5, 9, 10, 47, 50, 51, 56,
      61, 65

Vowden   68

$V_p(A, a)$ 20

$V_p(A; a_1, \ldots, a_n)$ 24

$V(T)$ 81

$v(T)$ 84

White   7, 87

Williams   6, 30, 88, 89

$W(T)$ 85

X (cartesian product)   87

X' 81

Yood   106, 107

Zelazko   32, 112, 116

Zenger   100, 126